Stone Wall Secrets

Exploring Geology in the Classroom

TEACHER'S GUIDE

Ruth Deike

Tilbury House, Publishers
Gardiner, Maine

Tilbury House, Publishers
132 Water Street
Gardiner, ME 04345

First printing: August, 1998
10 9 8 7 6 5 4 3 2 1

Text Design and Layout: Nina Medina, Basil Hill Graphics, Somerville, ME
Editorial and Production: Jennifer Elliott, Barbara Diamond, Mackenzie Dawson
Printing and Binding: InterCity Press, Rockland, Massachusetts

STONE WALL SECRETS

Kristine and Robert Thorson
Illustrated by Gustav Moore

Hardcover, $16.95; ISBN 0-88448-195-6
Grades 3–7

9 x 10, 40 pages, color illustrations
Children/Science

What can the rocks in old stone walls tell us about how the earth's crust was shaped, melted by volcanoes, carved by glaciers, and worn by weather? And what can they tell us about earlier people on the land and the first settlers? As Adam and his grandfather work together to repair the family farm's old stone walls, Adam learns how fascinating geology can be, and how the everyday landscape provides intriguing clues to the past. Gus Moore's richly detailed paintings are the perfect complement to a story full of imagery and wonder, a story that also shows positive family dynamics between different generations and different races in an adoptive family.

BECOME A ROCK DETECTIVE PROGRAM

The Rock Detective program is:

• A Teacher's Choice Award-winning education program based on the idea that students remember what they discover for themselves;

• Designed for any teaching setting. The program provides a hands-on approach with a fun and easy-to-teach format;

• Student tested! Young scientists cannot fail in this program and show increases in both knowledge and self-confidence. We often hear, "I want to be a scientist when I grow up";

• Made up of earth science activities that exceed National Science Education Content Standards recommended by the National Research Council;

• Utilized by such prestigious educational leaders as the Smithsonian Institute and endorsed by Sea World, Reading is Fundamental, Inc., the Association for Women Geoscientists, and many more;

• A 501(c)3 non-profit educational project whose mission is to encourage the teaching of pre-college earth science. All proceeds from the sale of Rock Detective materials support recovery programs for adults who were abused as children;

• Available as a customized, indestructible kit with complete instructions and large rock, fossil, and mineral samples.

For more information, contact The Rock Detective, 593 Gardiner Road, Dresden, ME 04342; phone 207-737-4612; fax 207-737-4031; e-mail kidsrx@agate.net; web site is http://www.rmid.com/rocks

Contents

Acknowledgments

I want to thank my students, particularly the little ones, third and fourth graders—always right there, and always with a question! Their bright minds are a light shining down the path—like they already know the way. The help of my colleagues with the U.S. Geological Survey and in my new State of Maine is gratefully acknowledged. And I thank the minions who have brought us the Internet. Without this fount of information at my fingertips I could never have done this work. My deepest thanks go to my daughter Kristen, who fed the animals and me; to Lauren Haven, who corrected *ad nauseam*, supplied most of the Internet references, and patiently handled panics of all sorts; and to Jennifer Elliott, my editor-mentor, who brilliantly guided things along.

I am grateful to several individuals who supplied me with faxed or verbal information, particularly Anne Stocking, of Ellsworth, Maine, for information on the rocks used by early New England Indians; and Robert Tilling of the U.S. Geological Survey for material on Mt. St. Helens and Hawaiian volcanism. There are many who helped directly or indirectly by making information available on the Internet. I would like to particularly mention Page Keeley, Science and Technology Specialist at the Maine Mathematics and Science Alliance (http://www.mmsa.org) and AMEYA, Archaeological Education Association, Londonderry, NH.

I want to especially acknowledge Kristine and Robert Thorson, the authors of *Stone Wall Secrets*, for the delightful way they portrayed the relationship between Grampa and Adam. As the old man leads his grandson along the stone walls like his father before him, we see the magic that teachers hope for in opening student eyes and minds to the wonder of things. I had to chuckle at the dreamy personality of Grampa, a venerable old geologist (and New Englander, to boot!). The factual description of the geological events Grampa could imagine gave me wonderful toeholds on which to base this guide. At Jennifer's suggestion, I simply asked the kinds of questions I would expect from Adam or my students, and then wrote the guide in order to answer as many questions as practical.

The manuscript in whole or in part has been read and greatly improved by several earth scientists: Dr. Sandra H. Clark and Jane Ferrigno, both geologists with the U.S. Geological Survey, Reston, VA; Dr. Richard Batt, Professor of Geology, Buffalo State College, NY; Dr. Robert Thorson, Professor of Geology and Geophysics, University of Connecticut at Storrs; James V. O'Connor, Geologist for the District of Columbia; Dr. Arthur M. Hussey, II, Professor of Geology, Bowdoin College, Brunswick, ME; Dr. William B. White, Professor of Geochemistry, and Andrew A. Sicree, Curator at the Earth and Mineral Sciences Museum—both at Pennsylvania State University, University Park, PA; Dr. Beverly D. Kay, Chair, Dept. of Land Resource Science, University of Guelph, Guelph, Ontario, Canada; and Dr. Carl B. Agee, Loeb Associate Professor of the Natural Sciences, Harvard University, Cambridge, MA. I thank them all.

Introduction

Grampa grinned. He reached for the small, white stone, lifting it high into the sunlight, where it sparkled in front of the boy's brown eyes. Then the old man began his story—a stone wall story—one that took them back through time.

—From *Stone Wall Secrets*

The charm of *Stone Wall Secrets* is that Grampa, just by being his dreamy self, becomes the door through which his grandson Adam can glimpse ancient beaches and woolly mammoths. Grampa is a wonderful teacher, so I am writing this as though Grampa were continuing his day with Adam. I can imagine that after lunch, on the front porch of their New England farmhouse, Grampa, with a faraway look, continues to answer questions that Adam had asked while his cider got warm. Adam's questions and those of my students are the keys for this book, opening doors into ideas and concepts more beautiful and wild than *Star Wars*, *Star Trek*, and *Superman* combined.

Like Grampa, and like most earth scientists, I, too, get a faraway look in my eyes when I think about time and how old the earth is—how it has been only an instant in the history of the universe that humankind has walked over rocks that tell us about ancient volcanoes and oceans. I got goosebumps when I first learned that the same earthquakes, landslides, and eruptions of hot volcanic ash that make headlines today have been happening for billions of years, and, as we shall see, that these very processes helped create the air we breathe and the lands we fight for.

How do we know these things? Because when they grow up, kids like Adam look for answers to questions they asked (or wanted to ask) while their cider got warm. Many of us have spent our lives trying to understand our beautiful planet. Curiosity is contagious, and my hope is that teacher and student alike will come down with the got-to-know-more's!

Why is it important to teach earth science at all? And why is it important to teach it so students will want to know more? I have strong thoughts on these subjects, and recently I sat in awe as the outgoing president of the National Association of Geoscience Teachers, Dr. Barbara Tewkesbury, gave a short, dynamic speech that rang with what I believe. Her talk was entitled "Today's Students—Do They Really Learn Differently?" She tells us why it's so important to hear our students' questions [italic type is my emphasis]:

On Saturday night, I took a cab in from the airport. My friendly taxi driver and I had a conversation that began predictably with the weather and veered in the direction of air pollution. At one point, he said, "I remember when I was little, I once asked my mother whether we would ever run out of air." I replied, "What a terrific question!" *I was thinking to myself what great questions kids ask about science and how it would be possible to fashion an entire course around that one question.* I then asked, "What did she say?" He replied, "It's funny—I remember it really well. She just laughed and laughed."

Here in this short conversation was a snapshot of what has happened to so many kids in science. They start out insatiably curious about the world, and they ask fantastic questions (after all, what is it about the earth's complex systems that means we don't run out of air?). In return, these budding young scientists, who come by their questions so naturally and so ingenuously, hear, in essence, "Well, if you knew enough about science, you wouldn't ask such a silly question." Science rapidly becomes for them a matter of not asking questions about the world, but of learning all of the complicated things that scientists have already figured out about the world....

• *Why does it matter? [Because] our students...will be...faced with issues* such as global warming, cloning, resource depletion, water quality, artificial intelligence, hazardous waste disposal, gene transplants, soil degradation, and maybe even off-planet colonies....

• *First, teach courses that reach all students.* The philosophy "if you can't run with the big dogs..." is the source of much student anxiety....

• *Teach geoscience as a way of finding out how the world works...science is the process of asking questions...and trying to find the best answers to those questions.... It's not the facts that excite scientists, but the process of discovery....*

• *Let students experience science.* No one would expect that a person who had sat in a lecture hall listening to someone talk about driving would become very skillful behind the wheel. Why should we expect that students who have listened to someone talk about science ought to be good at scientific reasoning? Students need firsthand experience and practice....

• *Build relevance into science courses.* Many students believe science is boring because they don't see how it connects to their own lives.... In addition, we need to consider building cultural relevance into our courses if we are interested in reaching more than just a few minority students, who commonly see geoscience as being a white, western, middle to upper class enterprise.

• *Help students see science as a creative and human enterprise....*
We urgently need to require all of our students to study more science and to urge our faculty to teach science more effectively in order to meet a challenge that is already upon us....

A great way to start studying earth science is to read *Stone Wall Secrets* aloud to your class and as you read, record the questions your students ask. You'll find that many will be answered in this Teacher's Guide because it was designed around the questions that kids like Adam, a typical ten year old, might ask. The questions fall into three major subject areas: Time, The Earth in Space, and Our Dynamic Earth. Within each major area, the questions group into several themes, such as The Origin of the Universe, The Rock Cycle, and What Happened to the Woolly Mammoth? I've provided introductory material for each section, using language and comparisons that will make it easier for you to explore these subjects with your students. The activities are designed so that students can discover answers for themselves, and resource lists provide suggestions for further reading and interesting Internet websites to visit.

A few notes:

We have provided four appendices. The Conversion Table will help you and your students work in both English and metric units. The glossary provides definitions in everyday language for many scientific terms and words. The references and a list of useful websites will help your students follow up on things that interest them. Each of the activities has been characterized with reference to the National Science Education Content Standards as delineated by the National Research Council in 1996 (see pages 108–113). A key to the content standards is shown below:

> K–4, 5–8 = Grade levels, U.S. Public School System
> Content Standards:
> A = Unifying Concept and Proceesses
> B = Science as Inquiry
> C = Physical Science
> D = Life Science
> E = Earth and Space Science
> F = Science and technology
> G = Science in Personal and Social Perspectives
> H = History and Nature of Science

Experts working in the varied fields of study touched upon in this guide may find some concepts to be oversimplified. For each theme, I sought out the important, central notions and explained them using analogies to things with which most students in third grade and up are familiar. My hope is that the concepts are not distorted in this process.

—Ruth Deike
Dresden, Maine

REFERENCES AND FURTHER READING:

National Research Council, 1996. NATIONAL SCIENCE EDUCATION STANDARDS: National Academy Press, Washington, DC, 262 pp.

Raymo, Chet and Maureen E. Raymo, 1989. WRITTEN IN STONE: A GEOLOGICAL HISTORY OF THE NORTHEASTERN UNITED STATES: The Globe Pequot Press, Old Saybrook, CT, 163 pp.

Rubin, Penni, and Eleanora Robbins, 1992. WHAT'S UNDER YOUR FEET?: U.S. Dept. of the Interior/U.S. Geological Survey, U.S. Government Printing Office, Washington, DC, 42 pp.

Stein, Sara, 1986. THE EVOLUTION BOOK: Workman Publishing, New York, NY, 389 pp.

Tewksbury, Barbara J., 1997. TODAY'S STUDENTS—DO THEY LEARN DIFFERENTLY?: National Association of Geoscience Teachers Outgoing Presidential Address, Geological Society of America Meeting, Salt Lake City, UT, October.

Watt, Fiona, 1991. PLANET EARTH, A PRACTICAL INTRODUCTION WITH PROJECTS & ACTIVITIES: Usborne Publishing Ltd., London, UK, 48 pp.

USEFUL WEBSITES:

National Education Standards Matrix, Berkeley, CA
•http://www.ucmp.berkeley.edu/fosrec/Matrix.html

Hundreds of Teaching & Learning Resources Across the Federal Government.
•http://www.ed.gov/free
"Federal Resources for Educational Excellence," (FREE).

TIME!

I-1
A Time Line

...the old man spoke about the sun, the moon, the stars, and how everything began.

—From *Stone Wall Secrets*

There was a time on the earth when there were no stone walls, no plants, no dogs or cats, no Christmas, and no baby brother or sister to steal our toys. In fact, we would not even be able to breathe on the early earth because there was no oxygen!

So how did we get here? Where did the air to breathe and the sun come from? And the stars to wish upon? How did anything get here? Was there a time of nothing? Brave and adventurous folks have asked these questions and then spent years and years looking for the answers. We may never have all of the answers, but there will always be young people who, like Adam, will begin to hear the great, woolly mammoths trumpeting across the glaciers, and they will get more and more curious. So come, let's sit beside Grampa on the front porch steps and ask about how it all began.

To anchor our wandering minds we have made a Chart of Major Events in the Geologic History of New England. (See below.) Now right off the bat we should talk a bit about geologic time. Grampa at age seventy or so would be considered a senior citizen, while Adam at ten years old is still close to the beginning of his life. While human time is measured in tens of years, geologic time is measured in *millions* (1,000,000) and *billions* (1,000,000,000) of years. In fact, look at the first event on the chart—the Beginning of the Universe. It happened some *fifteen thousand million* years ago—and that's *fifteen* (15) *billion* years. So geologic history is a special kind of history that goes all the way back to the very beginning of everything.

The story of how the universe with its twinkling stars came to be—The Big Bang Theory—is the first story we'll hear in Chapter II. And we'll also find out how scientists think the sun originated. Notice the birthday of our sun and fellow planets wasn't until the universe was almost 10 billion years old. So the earth was not here to see the beginning of it all!

But back to the early earth. It was a land without life. That's right, there wasn't anything that we would consider alive—no plants, no animals, not even germs. The next time you go to a shopping mall, or take a ride in the city, remember that once upon a time the earth had no life at all. It took hundreds of millions of years for the early earth to show signs of life. Notice on the chart that scientists believe "life" emerged on the earth about 3,850 billion years ago.

And what was this early life like? Well, there are no fossil remains of the very earliest life, so the answers must come from careful laboratory work. Many, many researchers have worked on this issue for most of the last two hundred years in laboratories all over the world, and the research will continue as long as there are scientists who ask this wonderful question. Their work shows that the oldest lifelike things were the outcome of a long evolutionary process to develop the building blocks of life (amino acids, enzymes, RNA, and DNA, etc.). There was no oxygen on the early earth, so these early life forms used other energy sources, possibly from minerals. The oldest life forms that left us fossil evidence look like bacteria—tiny, tiny little globs. And in 1996 it was very exciting to watch scientists disagree about similar-looking things that were found in a meteorite that probably came from Mars! Some researchers believe the "things" in the meteorite suggest life on Mars, and others argue just as strongly that they are not related to life at all!

How about the air we breathe: where did it come from? Scientists now pretty much agree that our atmosphere came from volcanoes, and that the oxygen in today's air was made over a long, long time by tiny little plant-like things called blue-green algae, or cyanobacteria (Kingdom *Monera*)—looking something like the slippery, green mossy stuff on rocks in many streams. These little organisms, zillions of them, use carbon dioxide and sunlight for food, and give off oxygen as a waste product! Look on the Chart of Major Events and see how old the earth was when its atmosphere began to contain significant amounts of oxygen.

I-2

A Chart of Some Major Events in the Geologic History of New England

(Events in **bold** are mentioned in *Stone Wall Secrets*)
(Continental collisions are **<u>underlined and bold</u>**

AGE, in MILLIONS of Years Before Present (Multiply each by 1,000,000)
("~" means approximately)

15-11,000	BEGINNING OF THE UNIVERSE
4,560	**Beginning of our solar system—birth of the sun and the planets**
4,550	**Age of Stony Meteorites (87% of all meteorites)**
4,500	Early earth may have been blasted by HUGE microplanet—moon formed from material thrown into space
4,500	PRECAMBRIAN GEOLOGIC TIME BEGINS
4,440	Age of Moon Rocks brought back by Apollo astronauts
3,990	Oldest dated rocks on earth
	• In North America, the Atasca Gneiss near Great Slave Lake, Canada
~3,850	**Emergence of life on earth**
	• Organic molecules evolve into the first cells
~3,800	ARCHAEAN GEOLOGIC EON BEGINS
3,800	Oldest sedimentary rocks—Greenland
3,500	Oldest known fossils—all are prokaryotes (DNA loose in cell)
	• Fossil bacteria from Australia, Warrawoona Group
	• Oldest known stromatolites (contain microbial communities including cyanobacteria), Archaean Pilbara Group, Australia
	• Cyanobacteria start producing oxygen

2,500	PROTEROZOIC GEOLOGIC EON BEGINS

2,300 **Beginning of transition to oxygen-rich atmosphere**
1500-500 Multicellular marine organisms evolve
1,400 Algae [eukaryotes (DNA inside nucleus)] abundant in oceans

1300-1100 Grenville metamorphism
    • Ancestral Adirondacks
800 Oldest fossils of protozoans of animal and plant affinities
700-500 Ediacara fauna (early fossils that resemble segmented worms, jellyfish, and sea pensfound today)
    • Shallow ocean environment

544 CAMBRIAN GEOLOGIC PERIOD BEGINS

544-505 Trilobites (extinct arthropods, which means "jointed foot," distant relatives of horseshoe crabs) dominant
544-500 **Iapetus Ocean covers New England**
    • Big landmass, called Avalonia, forms way out in Iapetus Ocean to the east

495 ORDOVICIAN GEOLOGIC PERIOD BEGINS

470 **Sulfur-smelling volcanic islands appear to the east of New England**

465 **Iapetus Ocean shrinking as ocean plate subducts to the east beneath the volcanic islands**
    • Muddy sediments spread over western New England

450 <u>**Taconic Mountain Building Cycle (Taconic Orogeny)**</u>
    • <u>Volcanic islands collide with North America</u>
    • Mountains made from the squeezed ocean floor of Iapetus Ocean
    • Slabs and slices of hardened muddy sediments (shale and slate), and beach sand (sandstone and quartzite), thrust upward into gigantic mountains. New England squeezed, hard! Rocks that will make Grampa's stone wall are munched and crunched.

440 SILURIAN GEOLOGIC PERIOD BEGINS

420 **Earliest land plants**

~420 **Rivers roar off the Taconic mountains and carry "peanut brittle" pebbles**
    • Pebbles and mud eventually end up as the brown, muddy sandstone conglomerate rock Adam and others find in their stone walls
    • Sandy beaches form along the coast of Iapetus Ocean. These beaches will harden, then erode into sandstone with sparkly quartz grains—the pieces of sandstone will appear in New England's stone walls.

415 DEVONIAN GEOLOGIC PERIOD BEGINS

    • Iapetus Ocean still shrinking—Avalonia landmass approaches New England

400 <u>**Acadian Orogeny**</u>
    • <u>Avalonia collides with New England</u> resulting in the Acadian Mountain Building Cycle

360 CARBONIFEROUS GEOLOGIC PERIOD BEGINS

375-335 <u>**Acadian Mountain Building Cycle continues**</u>

286 PERMIAN GEOLOGIC PERIOD BEGINS

280 <u>**Alleghenian Orogeny**</u>

- **<u>Gondwanaland (an ancient landmass including what is now Africa and South America) rams east coast of North America</u> resulting in the assembly of Pangaea**

245	Lots of plants and animals including the Trilobites go extinct
245	TRIASSIC GEOLOGIC PERIOD BEGINS
230	**Beginning of breakup of Pangaea**
~225	Dinosaurs appear
208	JURASSIC GEOLOGIC PERIOD BEGINS
~205	Dinosaurs leave footprints now seen on rocks in Connecticut
~162	Earliest birds *Archaeopteryx*
~160	Dinosaurs abundant, many large sauropod dinosaurs (long neck, long tail) e.g., *Apatosaurus*
146	CRETACEOUS GEOLOGIC PERIOD BEGINS
100	New large, fierce dinosaurs appear—*Tyrannosaurus rex*
~100	Time portrayed in movie *Jurassic Park*
65	****** THE CRETACEOUS-TERTIARY (K-T) BOUNDARY ******
65	TERTIARY GEOLOGIC PERIOD BEGINS
65	PALEOCENE EPOCH BEGINS
65	Big meteorite hits the earth near Chicxulub (chick-shu-lube) in the Yucatan peninsula in Mexico
65	Lots of plants and animals go extinct
65	Most dinosaurs extinct (especially the large, featherless ones)
54	EOCENE EPOCH BEGINS
38	OLIGOCENE EPOCH BEGINS
38	Mammals abundant
23	MIOCENE EPOCH BEGINS
5	PLIOCENE EPOCH BEGINS
3.5	African climate drier and cooler—woodland habitat of Australopithecines begins to disappear
3.2	LUCY, *Australopithecus* appears in Africa
~2.4	*Homo rudolfensis* appears in Africa. He was FIRST of our genus!
~2.4	*Australopithecus* disappears
1.8 to Present	QUATERNARY GEOLOGIC PERIOD BEGINS
1.8	PLEISTOCENE EPOCH BEGINS
~1.8	**Beginning of Great Ice Age (it is still in progress!)**
~1.8	*Homo rudolfensis* disappears
1.6	*Homo erectus* appears in Africa (Turkana boy)

AGE, in THOUSANDS of Years Before Present, (Multiply each by 1,000)

30	*Homo neanderthalensis* well established
~200	*Homo erectus* gone?
~130	*Homo sapiens sapiens* appears in Africa
~34	*Homo sapiens* migrates into Europe
~30	*Homo neanderthalensis* extinct
~20	**Age of the glacier that made grooves on the smooth black stone in Grampa's Wall**
~20	**Grampa's meteorite is picked up by a glacier**
20-18	**Age of the last glacial advance into New England**
~17	**Glacier dumps rock over land that would one day be Grampa's farm**
~17	**Gusty winds fill the air with find brown sand**

~17	Woolly mammoths trample the spongy green tundra in New England
~14-13	Last New England glacier melts
11 to Present	HOLOCENE EPOCH BEGINS
~12-11	First Paleo-Indians arrived in New England. They knew how to make fire.
~10	Most woolly mammoths gone
~12-9	Nomadic Paleo-Indians make long flint knives
~9-2.7	Archaics, the village dwellers
~5.0-3.8	Laurentian Stage, "Red Paint People"
~3.8-3.2	Susquehanna culture
~2.7-.35	Woodland Indians cook in fired clay pots
~.35 to Present	Contact Period, first contact of Native Americans and Europeans

AGE, in years *Before Present* (BP) [AGE, in years *After Christ* (AD, *Anno Domini*) in brackets]

| 350 BP [1650 AD] | Colonists from Europe clear the land and burn the trees |
| 200 BP to 50 BP [1800 to 1950 AD] | Grampa's farm is a busy place—the stone walls are built |

REFERENCES AND FURTHER READING:

Geologic ages are constantly being fine-tuned as dating technology advances and new areas are studied. An excellent reference to new Paleozoic chronology may be found in the following:

Tucker, R.D. and W.S. McKerrow, 1995. EARLY PALEOZOIC CHRONOLOGY: A REVIEW IN LIGHT OF NEW U-PB ZIRCON AGES FROM NEWFOUNDLAND AND BRITAIN: Canadian Journal of Earth Science, v. 32, p. 368-379.

Bengtson, S. (Ed.) 1994. EARLY LIFE ON EARTh: Nobel Symposium No. 84, Columbia Univ. Press, New York, NY.

Busbey III, A.B., and R. R. Coenraads, D. Roots, and P. Willis, 1996. ROCKS AND FOSSILS: The Nature Company Guides, Time-Life Books, Weldon Owen Pty Ltd, Sydney, Australia, 288 pp.

Harbaugh, John, 1970. STRATIGRAPHY AND GEOLOGIC TIME: WM. C. Brown Company Publishers, Dubuque, IA, 113 pp.

Melosh, H.J., (Ed.), 1997. ORIGINS OF PLANETS AND LIFE: Annual Reviews, Inc., Palo Alto, CA, 380 pp.

Press, Frank and Raymond Siever, 1998. UNDERSTANDING EARTH: W.H. Freeman and Company, New York, NY, 682 pp.

Raymo, Chet and Maureen E. Raymo, 1989. WRITTEN IN STONE, A GEOLOGICAL HISTORY OF THE NORTHEASTERN UNITED STATES: The Globe Pequot Press, Old Saybrook, CT, 163 pp.

Taylor, T.N. and E. L. Taylor, 1993. THE BIOLOGY AND EVOLUTION OF FOSSIL PLANTS: Prentice Hall, NJ. (Cyanobacteria Reference)

Thompson, Ida, 1996. NATIONAL AUDUBON SOCIETY FIELD GUIDE TO NORTH AMERICAN FOSSILS: Alfred A. Knopf, New York, NY, 846 pp.

USEFUL WEBSITES:

Excellent Site for Geologic Time Information
•http://www.ucmp.berkeley.edu/help/timeform.html
The University of California Museum of Paleontology.
Original Geologic Time Machine conceived and created by Allen Collins—11/26/94.
Revised format and concept by Robert Guralnick and Brian R. Speer—9/15/95.

Dinosaurs
•http://www.fmnh.org/exhibits/dino/Triassic.htm

On-line dinosaur exhibit at the Field Museum of Natural History, Chicago, IL.

Online Publication
•http://pubs.usgs.gov/gip/geotime/
The full text of the USGS publication titled *Geologic Time* is available online.

Lesson Plan for Teachers on Geologic Age Determination
•http://www.usgs.gov/education/learnweb/Lesson4.html
Using Radioactive Decay to Determine Geologic Age (for grades 7 - 12).

—SIDEBAR—

The Origin of Land Plants— and How Scientists Learn About Things

Many scientists have helped us understand beginnings, and Grampa is moved by what we have learned about the way things were so long ago on our earth. A geologist-paleontologist in Maine, along with several of his colleagues, is looking at the misty origin of land plants some 420,000,000 (four hundred twenty million) years ago during the Silurian Period. The first plants that moved ashore were probably moss-like forms, similar to marine plants like algae and seaweed. Most plants are photosynthetic organisms that use carbon dioxide and sunlight and produce oxygen and plant cells, and scientists need to be detectives in order to figure out what changes on the earth, and changes in the plants themselves made it possible for them to survive out of the ocean.

Plants couldn't survive on land until the ozone layer got strong enough to shield land from harmful ultraviolet (UV) light. UV light rays don't go through more than a few inches into seawater, so at first life was only possible in the ocean. The ozone layer forms in the upper atmosphere from oxygen, which on the early earth came from photosynthesis of blue-green, marine algae. And it took a long time before there was enough oxygen in the atmosphere to form an ozone layer!

Later, trees were possible when plant cells evolved the shape of long tubes. These long tube-like cells were important because land plants had to grow upward to get sunlight, and the cells had to move water up high off the ground (a soda straw does the same thing). These long cells also had to be strong, because they did not have the support of water around them like marine plants. Once cells like these developed, or evolved, plants could grow upward with long trunks and spreading branches.

Where do scientists search for these early plants? A great deal of such scientific detective work is begun in big libraries where the results of previous studies suggest areas to look for them. The researchers travel to the far reaches of the earth to look for fossil remains. Back in the laboratory, carefully collected samples are analyzed and compared with other finds from all over the world. Gradually, over months or years, the story comes into focus and a new beginning is understood.

I-3
Making a Time Line

OBJECTIVE: To help students understand that compared to a human life span, geologic time is HUGE.

A time line is a way of *picturing* some period of time. From the earliest, high energy beginnings of the universe to the present here on earth covers a span of some 15 billion (15,000 million) years. Here are a series of comparisons that help to understand this enormous period of time.

First, let's look at the size of one billion:
- One billion earths still would not equal the mass of the sun
- One billion minutes is 1,903 years
- One billion atoms make up the dot over this *i*

And the size of one million:
- If a person lived for one million days, he or she would be 2,740 years old!

ACTIVITY I
Making a Geologic Time Hall for Your School

OBJECTIVE: How long have humans been on the earth compared to the age of the universe? Or compared to the age of the earth?

> National Science Education Content Standard(s), K-4, 5-8
> Major Emphasis: A, D, E; Minor Emphasis: B, G, H;
> Other Content: Mathematics, Measurement Scale, Art

MATERIALS NEEDED: Tape measure; 3 x 5 cards; felt-tip markers of several colors; clear adhesive tape; calculator; blackboard and chalk

PROCEDURE:
1. Measure the length of the longest hall in your school. Use a hall that your class can safely go to during class time.
2. Have your class pick the ages of a series of events from the Chart of Geologic Events. (Note: Have the whole class pick the events as a group.)
3. Make three time lines and place the events on the lines.
   - Use 1/10 inch per 10 million years and place events on time line.
   - Use 1/10 inch per 1 million years.
   - Use 1/10 inch per 100 years.
4. Have your class draw a picture or other symbol to represent each event on the time line.

For each time line have the students answer the following: What is the oldest event on each time line?

The Age of the Universe as Distance Traveled

OBJECTIVE: To help visualize the age of the universe

> National Science Education Content Standard(s), K-4, 5-8
> Major Emphasis: A. E, H; Minor Emphasis: B;
> Other Content: Mathematics

MATERIALS NEEDED: A calculator; an encyclopedia; a globe of the earth

INFORMATION NEEDED:
- The age of the universe for this activity is 10,000,000,000 (ten billion) years. (Note: The more accurate age of the universe is between 11 and 15 billion years, but these are more difficult numbers to work with.)
- The circumference of the earth is approximately 24,000 miles (40,000 km).
- One mile equals 63,360 inches (see conversion table for metric units).

PROCEDURE:
1. Represent the age of the universe as a distance traveled. If you were so small that it takes one year to travel one inch, then since the universe was born you have traveled 10 billion inches.
2. Calculate how many miles equal 10 billion inches. We know that one mile contains 63,360 inches. Divide 10,000,000,000 (ten billion) inches by 63,360 inches per mile and you'll get 157,828 miles. That's the distance you could travel since the beginning of the Universe, if you went at the rate of one inch per year of geologic time!

     And how many times around the earth could you go? To answer this divide 157, 828 miles by the circumference of the earth, 24,000 miles.
   - The answer: Approximately 6 and one half times!

     What this means: Traveling at the rate of one inch per year, one could go six and one half times around the earth in the time since the universe was born!

REFERENCES AND FURTHER READING:

Berry, W., 1987. GROWTH OF A PREHISTORIC TIME SCALE: BASED ON ORGANIC EVOLUTION: Revised edition, Blackwell Scientific Publications, Palo Alto, CA, 202 pp.
Renne, Paul R., 1995. DATING EARTH'S MIDDLE AGES: Geotimes, v. 40, no. 11, November 1995 [geotimes@agi.umd.edu].

USEFUL WEBSITES:

Geo-Time Table
•http://www.m-w.com/mw/table/geologic.htm
Miriam Webster's On-Line Dictionary.

Dinosaurs
•http://www.fmnh.org/exhibits/dino/Triassic.htm
On-line dinosaur exhibit at the Field Museum of Natural History, Chicago, IL.

Geologic Time and Paleontology
•http://www.ucmp.berkeley.edu/fosrec/TimeScale.html
The Geologic Time Scale—Learning from the Fossil Record, a site from University of California, Berkeley, with lots of educational resources and activities.

When Did Time Start?

OBJECTIVE: To give students the chance to think about an important earth science issue—the beginning of time—and to make hypotheses (suggest answers).

Since there were no people around when time started, we can only hypothesize about it. That means we first ask a question and then suggest an idea, or hypothesis. Here are two questions we can use to practice making hypotheses:
1. Has there ever been a time of, "no time"? and,
2. Did time start when the universe was born?

There can be more than one hypothesis for any question. In fact, the more ideas we have, the better, because that gives us many approaches to solving the mystery.

(Mathematicians deal with this difficult question in a simplified way. They put a zero— "0"—on one side of their paper and infinity—"∞"—on the other side and pick a chunk of time out of the space in between. In other words, they can start things when they want.)

ACTIVITY I

To Determine Several Hypotheses from Class Discussion on the Question: Was There Ever a Time of No Time?

OBJECTIVE: To experience the process of making hypotheses (suggesting answers).

> National Science Education Content Standard(s), K-4, 5-8
> Major Emphasis: B, E, H; Minor Emphasis: A;
> Other Content: Language Arts, Public Speaking

MATERIALS NEEDED: Blackboard and chalk

PROCEDURE:
1. Divide your class into two groups (for interest, have them count off by one and two). Have all the "ones" go to one side of the room and the "twos" to the other side. Have the two groups speak to the question: "Was there ever a time of no time?"
2. To get things started, ask each group to gather together in a huddle and see if they can come up with a group opinion about the question in, say, ten minutes.
3. Now, have one student from each group speak alternately about the question until all students have had a chance to speak. Make sure all the students have a chance to suggest an answer. (Note: For students who don't seem to have anything to say, ask them to draw a picture, write a poem, or another way of expressing their answer.)
4. The students will end up providing a wealth of "hypotheses" for this interesting question. As the students give their ideas, help them make, or "formulate," an hypothesis, and list all the hypotheses on the board.
Example:
Here are two opposing hypotheses for the question, Was there ever a time of no time?:
1. There was never a time of no time because we use clocks to measure time, and without a clock time cannot be measured.

2. Time exists whether there is a clock to measure it or not—hence, there *was* a time before time.

Go through the list and see if the students can come up with ways to *test* each of the hypotheses. Keep in mind that some hypotheses will be untestable at the present level of technology. And some of the ideas will be unresolvable except by religious conviction—these, by definition, will not be hypotheses, but of course, they can still be debated.

Scientists use a similar method to arrive at testable hypotheses. Most questions will be much simpler than our questions about time. For example, with respect to growing plants, two hypotheses would be: 1) Plants that receive water once a month will die; and 2) Plants that receive water once a month will not die. These hypotheses can easily be tested.

If your class enjoyed this activity, have them do Activity II.

ACTIVITY II

To Determine Several Hypotheses from Class Discussion on the Question: Did Time Start When the Universe Was Born?

OBJECTIVE: To experience the process of making hypotheses (suggesting answers).

> National Science Education Content Standard(s), K-4, 5-8
> Major Emphasis: B, E, H; Minor Emphasis: A;
> Other Content: Language Arts, Public Speaking

PROCEDURE IS THE SAME AS ACTIVITY 1.
Note: Students should be encouraged to do additional work on poems, drawings, or written essays on these questions, and on ways of testing hypotheses. The results are guaranteed to be fascinating, and it is a wonderful way for them to learn about how science arrives at answers!

I-5

How Do We Measure Time?
Reading the Stones

OBJECTIVE: To help students understand the methods of determining the age of rocks based upon radioactive decay and relative dating based upon the fossil record.

> *Grampa stood up straight, strangely content. He saw that Adam was staring off into space,*
> *still and quiet, as if in a trance. Grampa wondered, "was I right to hope that he's ready to*
> *learn—to read the stones, to travel through time, to hear the earth as it speaks to me?"*
> From *Stone Wall Secrets*

When, like Adam, we pick up a rock, how do we know how old it is and how long it has been on its journey to us? We can answer these questions if our rock contains either a *fossil* that is of some known age, or a *mineral* that is radioactive, because radioactive minerals are nature's clocks.

Let's say we find a fossil in our rock. And we take our fossil to a museum and learn that it is the shell of a 450-million-year-old animal, similar to a clam, that lived in the mud in an ancient ocean. Well, that's pretty exciting! Imagine, a little clam-like animal that is millions of years old! But then we get to wondering, how does our helpful museum scientist know how old this little guy is? She tells us that our little animal has been found in many places all over the world, and rocks containing these fossils have been "dated" with minerals that are radioactive! The plot thickens.

Our museum scientist explains that hundreds of millions of years ago our shell was buried along with a radioactive mineral clock. Elements in the mineral clock were silently measuring the passage of time. These clam-like animals lived in bays and along ocean shores in many parts of the world. When our animal died, his shell—along with the shells of zillions of his cousins, brothers, sisters, aunts, and uncles—became part of thin layers of mud in coastal areas near many continents. Eventually the muddy sediments turned to rock, and all the while the radioactive clock was marking time.

About 100 years ago, scientists discovered radioactivity in the elements uranium and radium. They found that radioactive elements pop out (or emit) a little burst of energy—sort of like turning a light bulb on for an instant. The little bursts of energy are like baseballs pitched from a bionic pitcher—they go by so fast we couldn't possibly see them, and some of them are so strong they would go right through the catcher! (Don't worry, the catcher wouldn't feel a thing!) But if we put an unexposed film behind the catcher we would see his or her bones because we have just taken an x-ray picture. Does this mean that the catcher is now radioactive and will glow in the dark? No, it doesn't, but the effect of radioactive particles (there are several) on the human body is an important question.

The scientists studying these elements found something else altogether fascinating. As the little bursts of energy are released, a radioactive element can change into a *totally different chemical element*! Uranium, for instance, becomes lead (fishing sinkers are made of lead). The new element is often stable and not radioactive. This discovery opened the door to radiometric dating, (using the radioactivity of naturally occurring elements to determine the ages of rocks) and then to the *age of the earth*! Here's how it happened: In 1905 Ernest Rutherford, a New Zealander working in England, used radioactivity to tell the age of a uranium-bearing mineral. To his surprise, the measurements indicated that the mineral was very old. This also meant that the rock, and indeed, the earth, must be very old! In the next few years, more rocks were dated and methods were improved, and soon it became evident that, in fact, the earth was billions of years old! Imagine Dr. Rutherford telling his family that he had just found that the earth was not thousands, but billions of years old!

How does a radioactive clock work? Well, the "baseball" from our bionic pitcher is actually one of several tiny, rambunctious (very energetic) particles from inside the element. Radioactive elements contain a few atoms that try to carry around too many particles, and at random intervals they drop the extra baggage. The escaping particle creates the burst of energy. The radioactive atoms in an element will steadily give off little bursts of energy until the atoms with extra particles are all gone! This is called radioactive decay. Over time, our radioactive element (called the parent) gradually changes into something different (called the daughter). It's like popcorn—if you think of the parent as unpopped corn kernels, and the daughter as "popped" kernels, then like popcorn popping—each time a corn kernel (the parent) "pops" there is one less parent left, and one more daughter. With very sensitive instruments, geologists count the number of parent atoms and the number of daughter atoms, and if there are only a few parents compared to daughters, then the rock containing the radioactive element has been "decaying" for a long time. And this is how the age of the radioactive mineral buried along with our little clam-like animal was determined.

I can hear the questions—what is an atom, and how does an atom get too many particles?

The Carbon-14 Clock

To answer these questions, let's take a look at carbon (C), a very popular element in Mother Nature's pantry. Along the way we can learn about the carbon-14 clock for dating relatively recent (less than about 70,000 years old) wood, bone, or anything else that contains the element carbon (which, by the way, includes carbon dioxide [CO_2] gas!).

First, what's an atom? If we were to cut a piece of charcoal, made of the element carbon, in half over and over again, the smallest piece we would end up with that would still be charcoal would be an *atom* of carbon. The atom would have some electrons orbiting like satellites around a center made of neutrons and protons. Radioactive atoms of carbon try to carry around one too many particles. These atoms are called carbon-14 for the number of neutrons and protons they have to carry around (chemists write carbon-14 this way: ^{14}C). How does a carbon atom get the extra particle? The most likely way is for some very, very strong force to "glue" the particle into the atom. For radioactive carbon, this happens out in space when cosmic particles (neutrons) whizzing about collide with *nitrogen atoms* that are minding their own business! These cosmic rays transform nitrogen atoms into radioactive carbon atoms. (Older students will enjoy taking a look at carbon and nitrogen on a Periodic Chart of the Chemical Elements. They will discover that the two elements are next to each other, meaning they differ only in the number of neutrons and protons in their respective nuclei, which is why, under the right conditions, it is possible for them to change back and forth.)

Most carbon atoms don't have extra particles to worry about. But about one in every 1,000,000,000,000 (one trillion) carbon atoms is radioactive—it will drop its baggage and emit energy. When the extra particle leaves, the whole carbon atom changes back into a nitrogen atom and disappears into the air. It is very interesting that compared to the number of regular carbon atoms, the number of carbon-14 atoms in our atmosphere stays pretty much the same. The cosmic crashes and escaping particles have been going on for billions of years, so the number of atoms with extra baggage is pretty constant. This is important because it is the starting point for the carbon-14 radioactive clock.

In order to use the radioactive carbon as a clock, it needs to become part of something on earth that we want to date, for example, the firewood from ancient Indian campfires. Firewood, of course, comes from trees, and trees are made mostly of carbon. Their carbon comes from the carbon dioxide (CO_2) in our atmosphere. Some of the CO_2 forms when carbon atoms, including ^{14}C (carbon-14) combine with oxygen. Trees breathe in the CO_2 and turn it into wood. As soon as the carbon atoms are tucked into the tree, the unfortunate ones that were turned into carbon-14 before the tree breathed them in, will continue to emit their extra baggage, and change into nitrogen (the "daughter") which will escape back into the atmosphere. Because the daughter, nitrogen, doesn't stay put, geologists must compare the amounts of carbon-14 and regular carbon *originally* found in the atmosphere to the amounts of carbon-14 and regular carbon left in the sample to find the age of the tree, or charcoal. Activity I will help students understand how this is done.

As you can imagine, some radioactive elements pop out energy faster than others. Atoms of the fast ones are all gone sooner than the slow ones. For this reason there are several elements needed to determine various earth ages. The element carbon is very useful for carbon-14 dating of charcoal in Indian campfires, but carbon-14 changes pretty quickly into the element nitrogen (N), and there is too little left to measure after about 70,000 years. On the other hand, uranium-238 (^{238}U), which changes to lead (Pb), takes a long time between energy bursts and can be used to date things as old as nearly 5 billion years!

The Class "Becomes" Two Radiometric Clocks

OBJECTIVES: To introduce the way radioactive decay of uranium-238 and carbon-14 are used to measure time.
  Activity I(A): The concept of half-life
  Activity I(B): Calculate the age of a "sample"

> National Science Education Content Standard(s), K-4, 5-8
> Major Emphasis: A. C. F; Minor Emphasis: B; Other: Math

INTRODUCTION:

The reason radiometric dating works is that the average rate of radioactive decay is fixed—an example of a fixed rate would be a car driving down the road at exactly the same speed forever. During decay parent atoms steadily change into daughter products.

Radioactive decay is expressed in terms of the radioactive element's half-life—the time required for one-half of the original number of radioactive atoms to decay. At the end of the first half-life after a radioactive element becomes part of a new mineral, half the number of parent atoms remain. At the end of a second half-life, half of that half, or one-quarter of the original number, are left. At the end of a third half-life, an eighth remains, and so on.

If we know the decay half-life of an element, and can count the number of newly formed daughter atoms as well as the remaining parent atoms, we can calculate the time since the radioactive clock began to tick. In effect, we can work back to the time when there were no daughter atoms, only those of the undecayed parent atoms. (For more information on how these measurements are made, see F. Press and R. Siever, 1998.)

In Activity I(A) we will use make-believe half-life times for uranium and carbon to demonstrate how the idea of half-life works.

In Activity I(B) we will use real half-life times (see Table I-5-1) to determine the age of samples of uranium and carbon bearing minerals.

ACTIVITY I(A)

Demonstrate the Concept of Half-Life

MATERIALS NEEDED: Two stopwatches (or a wall clock with a second hand) and a blackboard

PROCEDURE:
1. Divide class into two groups of students who will become atoms of "radioactive" elements. Group One will be Uranium 238 (^{238}U). Group Two will be Carbon-14 (^{14}C). The number of students in each group must be a "power of 2"— that is, 2, 4, 8, 16, 32, 64 etc. The two groups should each take one-half of the classroom.
2. Have each group name a timekeeper and give each a stopwatch. Have both groups stand up, and have the timekeeper count the total number of students in their group. Write the totals under "Uranium 238" and "Carbon-14" on the blackboard. Note: This number represents the number of undecayed parent atoms in your sample—you will use it later.

3. In Group One: ^{238}U decays to lead. So every 60 seconds the timekeeper tells one-half of the kids to sit down. Sixty seconds is the half-life we will use for our demonstration, the REAL half-life is 4.5 billion years!! So you can see how slowly uranium-238 decays.
4. In Group Two: ^{14}C decays to nitrogen. So every 10 seconds the timekeeper tells one-half of the kids to sit down. The REAL half-life of carbon-14 is 5,730 years, much faster than uranium-238.
5. Continue the "radioactive decay" until only one "atom" is left and record the total time.

The total times for each group of atoms to "decay" if you start with, say, 16 students, would be 240 seconds for the Uranium Group and 40 seconds for the Carbon Group. This shows the relative differences in the ages that can be measured by these two elements. (See Table I-5-1 below.)

Table 1-5-1 Two Major Radioactive Elements Used in Radiometric Dating

Element		Half-Life of Parent (Years)	Effective Dating Range (Years)	Minerals and other Materials That Can Be Dated
Parent	Daughter			
Uranium-238	Lead-206	4.5 billion	10 million- 4.6 billion	Zircon Uraninite
Carbon-14	Nitrogen	5730	100-70,000	• Wood, charcoal, peat • Bone and tissue • Shells • Groundwater, ocean water and glacier ice containing carbon dioxide

ACTIVITY I(B)

Now Let's Measure the Numbers of Parents and Daughters and Calculate a Real Age for a "Sample"

PROCEDURE:

1. Explain to your students that you are running a Radiometric Dating Laboratory and they will be the radioactive atoms decaying in a "sample" of a radioactive mineral for you to date. It is important that each student realize he or she is going to pretend to be either a radioactive atom or its daughter product for this activity.

2. Divide the class into two groups—one will be the radioactive uranium atoms in a sample of uraninite (a mineral containing uranium) from Canada; and the other group will be radioactive carbon atoms in a sample of charcoal from an Indian campfire site found in Connecticut.

3. Have the timekeepers for each group count the total number of students and write this number on the board; label it "Total Number of Parents."

4. Now explain to the groups that in each case, some time has passed since the minerals were formed, and that one-half of the radioactive atoms have decayed. Have the timekeepers ask every other student, or one-half of the group to "decay" and stand up. Note: if there is an odd number of students, have the timekeeper join the group. The standing students are the "daughter products"—count them and write that number on the board. Label it, "Number of Daughters."

5. Compare the numbers of parents and daughters for each group (depending upon the level of mathematics proficiency, you could divide the number of daughters by parents, or do a ratio) and note for your students that since one-half of the "parents" decayed to "daughters," that means one half-life has passed. Ask the BIG QUESTION: How old are the samples of uraninite and charcoal?

6. While the students are thinking about the answer, write on the board for each element:

Half-life of uranium-238 = 4.5 billion years

Half-life of carbon-14 = 5,730 years

7. Help the two groups to see that the ages of their samples are 4.5 billion years for the uraninite, and 5,730 years for the charcoal.

REFERENCES AND FURTHER READING:

Edwards, Lucy and John Pojeta, 1993. FOSSILS, ROCKS, AND TIME: U.S. Dept. of the Interior/U.S. Geological Survey, U.S. Government Printing Office, Washington, DC, 24 pp.

Harbaugh, John, 1970. STRATIGRAPHY AND GEOLOGIC TIME: WM. C. Brown Company Publishers, Dubuque, IA, 113 pp.

Newman, William, 1988. GEOLOGIC TIME: U.S. Dept. of the Interior/U.S. Geological Survey, U.S. Government Printing Office, Washington, DC, 20 pp.

USEFUL WEBSITES:

Using Carbon-14

•http://www.blm.gov/education/mesas/clues.html

A science detective story about a possible new prehistoric culture in Alaska—with some good descriptions of how carbon-14 dating is used.

Age-Dating Activity

•http://www.ucmp.berkeley.edu/fosrec/McKinney.html

 Uses M & M's!.University of California at Berkeley, CA.

Geologic Time and Paleontology

•http://www.ucmp.berkeley.edu/fosrec/TimeScale.html

The Geologic Time Scale—Learning from the Fossil Record, a site from University of California, Berkeley, with lots of educational resources and activities.

THE EARTH IN SPACE

II-1
Origin of the Universe

OBJECTIVE: To learn about the Big Bang Theory for the origin of the universe.

> *That night his sister pointed at a distant spot in the sky. She told him that he'd been born out there and then brought to earth by a falling star. He was so young then, he almost believed her. In fact, she insisted he owed his life to the shooting stars. Later, much later, he learned that his sister's ideas weren't really so far-fetched, that every atom in his body had first arrived on earth that same way.*
>
> *He explained this all to Adam. Keeping his eyes closed to protect them from the sunlight, the old man spoke about the sun, the moon, the stars, and how everything all began.*
>
> From *Stone Wall Secrets*

The idea that before our world as we know it there was *nothing*, is pretty scary, but at the same time it is exciting to wonder about. Famous people like Albert Einstein (1879-1955), Henrietta Leavitt (1868-1921), Edwin Hubble (1889-1953), and Stephen Hawking have worked very hard to learn about the origin of the universe.

The Big Bang
Out of their work has come a concept called "The Big Bang." This idea has stood the test of scientific debate since the 1940s when it was proposed by the Russian-American physicist George Gamow and his co-workers. The Big Bang Theory explains how atoms that make up our bodies were formed, and how out of nothing the universe came to be. Here's a synopsis:

The First Second: At the very, very beginning of it all, a tiny speck of brilliant light appeared. It was much hotter than our sun. All of space, everything we know was inside this little ball of fire. As you can imagine, it was pretty crowded in there! It was full of zillions of tiny particles of light and radiation all zooming around bumping into each other. There was so much energy in such a tiny space that "lumps" of matter appeared and disappeared. Just like *Star Trek!*

Suddenly, the infant universe *blew up!* Like a balloon that has had too much air—the tiny fireball blew apart, and within a fraction of time almost too small to measure it grew a hundred trillion trillion trillion trillion times, and like an air conditioner, it got very cold! The blow-up is called *cosmic inflation*, and it explains why stars and galaxies in the sky are evenly distributed, and why the universe is so big. During this flashing moment, the force of gravity separated itself (without gravity nothing in the universe would stay put!). The cosmic inflation also provided the tremendous force (called the "strong force") that would make it possible for matter to form. The strong force is what holds neutrons and protons together inside the atoms.

Almost as soon as it began, the inflation was over, and now particles that would become planets and people started to form. The young universe, still *less than one second old*, was in total turmoil. The temperature went back up, and in a tremendous surge of energy, the *particles* appeared. Physicists have named them gravitons, gluons, photons, bosons, neutrinos, WIMPs (weakly interacting massive particles), quarks, and leptons. Atoms are made of quarks and leptons, and atoms are the building blocks of all matter (trees, dogs, fire hydrants, and everything else). But as we will see, these particles had to fight for their survival.

For the rest of its first second of existence, the universe was a fierce battleground from which only a few types of particles survived. All of the particles were locked in a battle with antiparticles exactly like them. Most were annihilated. Our world today is possible because a tiny imbalance favored quarks and leptons over antiquarks and antileptons. Atoms would follow.

At the end of its first second, the boisterous infant universe was an incredibly huge, bright thing, because most of the particles that survived the annihilation were photons, which are tiny bundles of light. But this early universe would not have looked bright at all, because none of the light could get out! Photons and electrons collided with each other, keeping the photons from traveling very far, and light couldn't get through, so the universe was opaque. Like your car headlights shining into very dense fog, the early universe was impossible to see through. Physicists believe they can actually see this white opaque early universe! It is called the "afterglow," and we will find out more about it below.

From One Second to Three Minutes: During this time between one second and three minutes all the ingredients were available for cooking up the nuclei of the very *first elements*, hydrogen, helium, and lithium. When you buy a party balloon that floats in air, it is filled with helium gas created when the universe was only three minutes old!

> Calculations say that elements created in the early universe should be 77% hydrogen, 23% helium and tiny amounts of lithium. Gas clouds far out in space contain just these amounts of these elements—powerful evidence for the Big Bang Theory.

As soon as the elements took form, the universe became crystal clear, like it is today. The zillions of pesky electrons were pulled into orbit around the nuclei (center) of the hydrogen and helium atoms. The electrons no longer banged into the passing photons; light had a free passage—and space became transparent. Scientists today who look at the oldest light from the farthest reaches of space see a wall of white fog. The most powerful telescopes can see to this wall and no farther; most scientists believe this is the afterglow predicted by George Gamow, and that the "white wall" clinches the Big Bang Theory.

Three Minutes to 300,000 Years: After the busy first three minutes, the universe settled down in a much calmer period that lasted more than 300,000 years during which the stage was set for galaxies like our Milky Way to form. Gradually, with the new force of gravity, the ingredients of the cosmos started pulling on each other. Like milk curdling into cheese, gases made mostly of helium and hydrogen were pulled into denser regions. Heat is used up trying to escape from these more dense, ancient gaseous regions and compared to the rest of space, the curdled regions appear cooler. Our evidence for this part of the story came in 1992 when COBE, the Cosmic Background Explorer, detected tiny differences in the temperature of the cosmos. Astronomers had been looking for the raw material for building galaxies, and COBE found it! These cooler curdled lumps of gaseous matter provided the material for galaxies to grow.

How Old is the Universe?

Because most scientists agree that the Big Bang Theory explains things quite well, the focus of research has shifted from *how* the universe formed to *when*. In fact, the main task of the Hubble Space Telescope is to provide information that can be used to measure the age of the universe. Scientists use three very different ways to estimate the age of the universe, and all three give similar ages. The age from decaying radioactive elements in some meteorites yields an average of 15 billion years. In the Milky Way galaxy, stars born soon after the universe itself average 14 billion years old. "Winding back" the expansion of the universe to find out when it started expanding provides a time span of 11 billion years. Averaging the three methods gives an approximate age for the universe of 13 billion years.

The Far Future

So where is it all heading? Well, scientists calculate that billions of years in the future the universe may literally disappear into its own mega (huge) black hole; or, some even have suggested that the proton, the building block of matter, may decay after trillions and trillions of years.

Other Big Bangs

If our universe came into being, why not other universes? One theory predicts that the matter disappearing down a black hole may "bud" off the bottom of the well to produce another universe!

ACTIVITY I

Class Presentation on "The Origin of the Universe"!

OBJECTIVE: To give students the opportunity to understand the main parts of the Big Bang Theory.

> National Science Education Content Standard(s), K-4, 5-8
> Major Emphasis: A. C. E; Minor Emphasis: B, H;
> Other Content: Language Arts, Public Speaking, Theater, Art

MATERIALS NEEDED: Large meeting room like the auditorium; about two weeks of class time; THE BIG BANG by Couper and Henbest (see below) is highly recommended as a superb resource.

PROCEDURE:

Have your class give a presentation to the rest of the school in the auditorium about the origin of the universe. Have a group of students each paint/draw and/or explain one part of the Big Bang Theory. To add interest, your students could create a play called "The Big Bang." Each part of the theory (see below) can be a quick scene illustrated with drawings and narrated briefly by a student. Some of the students could dress up as quarks, photons, electrons, leptons, etc. The battle scene between particles and antiparticles could be awesome!

The First Second
- A tiny speck of hot, brilliant light appeared.
- Such a tiny space that "lumps" of matter appeared and disappeared.
- Suddenly, the infant universe blew up!
- It grew a hundred trillion trillion trillion trillion times.
- It got very cold!
- The blow-up is called *cosmic inflation*, and it explains why the universe is so smooth and so big.
- The force of gravity separated itself (without gravity nothing in the universe would stay put!).
- The cosmic inflation also provided the strong force that holds matter together inside the atoms.
- The temperature went back up, and in a tremendous surge of energy, the particles appeared (gravitons, gluons, photons, bosons, neutrinos, WIMPs (weakly interacting massive particles), quarks, and leptons).
- The universe was a fierce battleground where particles were locked in a battle with antiparticles exactly like them. Most were annihilated. Ones to survive are photons.
- A tiny imbalance favored quarks and leptons over antiquarks and antileptons.
- Atoms are formed.
- At the end of its first second, the boisterous infant universe was an incredibly huge bright thing, but like the Klingon "Cloaking Device" (*Star Trek*), it was invisible!

From One Second to Three Minutes
- Atoms of the very first elements—hydrogen, helium, and lithium—appeared.
- The universe became crystal clear, like it is today.

Three Minutes to 300,000 Years
- The ingredients of the cosmos started pulling on each other.
- Gases made mostly of helium and hydrogen were pulled into denser regions.
- Galaxies grow from curdled lumps of gaseous matter.

REFERENCES AND FURTHER READING:

Couper, Heather, and Nigel Henbest, 1997. BIG BANG, THE STORY OF THE UNIVERSE: DK Publishing, Inc., New York, NY, 45 pp.

Fahs, Sophia Lyon, and Dorothy Spoerl, 1960. BEGINNINGS: EARTH, SKY, LIFE, DEATH: Beacon Press, Boston, MA, 217 pp.

USEFUL WEBSITES:

Space Education
•http://spacelink.nasa.gov/.index.html
NASA provides information and links for teaching about space.

II-2
Origin of our Piece of Space, the Milky Way Galaxy

OBJECTIVE: To help students understand present theories about the formation of our galaxy and our solar system.

If you could jump into your spaceship and zoom out into the stars, as you got closer you would find that many of them are actually millions and millions of stars in groups called galaxies. Many of the galaxies are in the form of spirals, looking like stirred paint or cake batter! Our galaxy, the Milky Way, is a beautiful spiral of stars that is 100,000 light years across. A light year is the distance light can travel in one year—that distance is about 5.9 trillion miles (9.5 trillion kilometers). Our Milky Way galaxy is shaped like a disk, sort of like a humongous pinwheel, (you can see a galaxy similar to ours in Figure II-2-1); the "wheels" are bent arms containing millions of stars. Our family of planets and our sun lie comfortably in one spiral arm, about halfway from the center to the edge of the galaxy. (Figure II-2-2 is a very special computer picture that shows our galaxy as though we were looking down on it from far out in space. You can see right where we live!) On the next clear night, go outside and look for the Milky Way; when you find it, you are looking at the edge of our galaxy!

Astronomers believe that our Milky Way galaxy, like most others, had a violent beginning. Gas clouds were pulled into the center and flared off a dazzling quasar; this sent jets of radiation far out into space, and our galaxy spun like a top on a spindle of intense light. At the center was a black hole, voraciously gobbling up new stars as quickly as they formed, and shooting what it didn't eat far out into space.

After a few million years our galaxy became less violent—the quasars evolved into jets of enormous clouds generating powerful radio waves (like the static you hear on your radio between stations!). As cosmic gas was used up to make stars, the black hole at the center was slowly starved.

It took nine billion years for our galaxy to "grow up" and become the spiral we live in today. During this time within the Milky Way, stars were born in huge pillars of dust and died in explosions that blasted heavy elements of iron and silica across the sky. These are the elements that would make our sun and planets. So you see we are truly born of the stars!

Figure II-2-1 The NGC 2997 Spiral galaxy, similar in shape to our Milky Way galaxy.
Source: Copyright Anglo-Australian Observatory, photograph by David Malin.
http://asca.gsfc.nasa.gov/docs/StarChild/universe_level2/milky_way.html

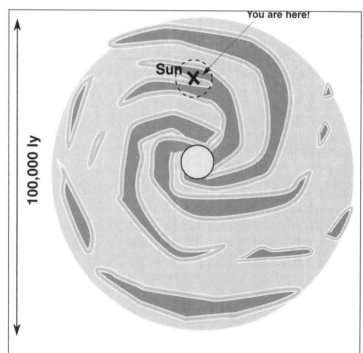

Top View of the Milky Way Galaxy

Hot, blue stars delineate spiral structure. Since hot stars are so luminous, they make the spiral arms stand out. Cool, orange and red stars are found in and between the spiral arms. The Sun's location in the Galaxy is marked. We are not at the center! Interstellar dust limits our view in visible light to roughly the area within the dashed circle around the Sun.

Figure II-2-2 Location of the Earth in the Milky Way galaxy
Source: http://www.bc.cc.ca.us/programs/sea/Astronomy/ismnotes/ismglxyb.htm
Copyright 1998 Dr. Nick Strobel. Used by permission.

We Can Actually See Our Galaxy, the Milky Way Galaxy

OBJECTIVE: To show the relationship between the earth and the Milky Way galaxy.

> National Science Education Content Standard(s), K-4, 5-8
> Major Emphasis: A, E; Minor: B, C, H
> Other Content: Art

MATERIALS NEEDED:

Sheet of posterboard at least 36 inches on a side
Large Styrofoam ball about 8 inches in diameter (available in the crafts department)
Ping-pong ball
Hot glue gun (works the best) or another type of glue

PROCEDURE:

1. Using the drawing in Figure II-2-2, enlarge and draw the Milky Way galaxy on the piece of poster-board.
2. Cut the Styrofoam ball in half and mount each half in the center of the "galaxy" on each side of the posterboard.
3. Cut the ping-pong ball in half and mount each half on the model where the solar system is shown on the diagram (this is not to scale!).
4. Now pick up the posterboard model and view it from the edge. It represents what your students will see at night when they look for the Milky Way.
5. On a dark clear night, go outside and look for a "milky" band of stars—that's the edge of our galaxy!

USEFUL WEBSITES:

Milky Way Galaxy: Cambridge University, England
•http://www.damtp.cam.ac.uk/user/gr/public/gal_milky.html

Virginia L. Murray Elementary School
•http://pen1.pen.k12.va.us/Anthology/Div/Albemarle/Schools/MurrayElem/ClassPages/Butl
er/SPACE/THEMILKYWAY.HTML

THE COSMOS
•http://csep10.phys.utk.edu/guidry/violence/ginfo1.html
Science and Engineering Research Council/Royal Greenwich Observatory .
Violence in the Cosmos—Explosive Processes and the Evolution of the Universe: Includes subjects such as
The Mother of All Explosions, The Stars: Some Don't Go Gently into the Night, The Galaxies: Some are
Quiet, Some are Not, and more.

Cambridge Cosmology: Galaxies
•http://www.damtp.cam.ac.uk/user/gr/public/gal_milky.html

Fun Facts for Kids about Space Stuff
•http://asca.gsfc.nasa.gov/docs/StarChild/universe_level2/milky_way.html
StarChild Learning Center for Young Astronomers.

Modern Cosmology
•http://map.gsfc.nasa.gov/html/milky_way.html
Microwave Anisotropy Probe: This web site introduces basic concepts in modern cosmology and describes
the MAP mission at a general level. (*Anisotropy* means not evenly distributed.)

Out of the Dust!
The Origin of Our Sun

OBJECTIVE: To demonstrate the forces that formed our galaxy.

When a twirling figure skater brings her arms close, she spins so fast she seems to disappear into a glittering solid (See Figure II-3-1). The same forces that increase the speed of the skater's spin brought galactic gas and dust clouds together and formed our sun! By the time our sun was formed, about 4.6 billion years ago, the Milky Way galaxy was filled with clouds of stellar gas and dust from the birth of the universe and from the birth and death of stars. Astronomers have found that the gas is made of the elements hydrogen and helium, and the dust particles are mostly siliceous minerals (similar to the composition of window glass), iron oxides (like magnetite, for example), and, ice crystals (yes, there is water in outer space!). These dust clouds usually slowly expand in the vacuum of space, but sometimes they get pulled from one side, and start to rotate. After the dust cloud starts to spin, the center portion gets heavier like the figure skater pulling in her arms, and the cloud will spin even faster.

Now a strange thing happens. The dust in the cloud starts to be attracted to the center of the cloud, which makes it even heavier and it spins even faster. Pretty soon, the center is so heavy that the dust particles press into each other with incredible force; in fact, some of the dust particles actually join together. Each time this happens a little energy is given off and pretty soon the dust cloud becomes a radiant star like our sun.

How long does it take to make a sun? Well, astronomers tell us that the universe began about 15 billion years ago, and meteorites tell us that the solar system was completely formed by 4.56 billion years ago. That means the sun could have taken about 10 billion years to form.

ACTIVITY I

Spinning

National Science Education Content Standard(s), K-4, 5-8
Major Emphasis: A, C; Minor Emphasis: B;
Other Content: Language Arts, Public Speaking

PROCEDURE:
1. Observe a figure-skating program on television.
2. Write down what you see when a skater spins and brings her arms closer to her body.
3. Watch several skaters (men and women) bring their arms in during a spin and see if the same thing happens to them all.
4. Be ready to give a report to your class.

A QUESTION FOR FURTHER STUDENT RESEARCH:
Question: How much would you weigh on the sun?
Answer: If you weigh 100 lbs now, you would weigh 2,800 lbs, or over one ton on the surface of the sun!

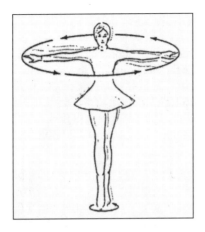

Figure II-3-1 Illustration of conservation of angular momentum; when a skater pulls in her arms, she spins at a faster rate. Similarly, when a slowly rotating nebula contracts, its rotation speed increases.
Source: F. Press and R. Siever, Earth, p. 7, Copyright 1978 W. H. Freeman and Co.

USEFUL WEBSITES:

Information about the Sun
•http://www.seds.org/billa/tnp/sol.html

StarChild: A Learning Center for Young Astronomers
•http://asca.gsfc.nasa.gov/docs/StarChild/universe_level2/milky_way.html
•http://map.gsfc.nasa.gov/html/milky_way.html

II-4

Clues from Meteorites About the Origin of the Earth

OBJECTIVE: To show that meteorites play a very important role in helping us understand the origin of the earth.

> *...the old man spoke about the sun, the moon, the stars, and how everything all began. How fragments like their meteorite collided and crashed and collected in space until all the planets were born. How leftover pieces still in orbit will fall for eons to come....*
>
> From *Stone Wall Secrets*

Like scientific gifts, meteorites come to us from outer space. They are what's left of meteors that crash to earth. Meteorites are made of the siliceous and iron-rich stuff of ancient dust clouds; they consist of the ashes of stars that have lived and died since the universe was born. Most meteorites that reach the earth come from within our solar system. Most are fragments, called meteoroids from the asteroid belt between the orbits of Mars and Jupiter (you can see the Asteroid Belt in Figure II-4-1).

Meteoroids are captured by earth's gravity and blaze down through the atmosphere. They are the meteors, or "shooting stars" that Grampa and his sister saw that night from their field by the pond. Some meteorites may have come from our moon, and scientists believe some have come from the planet Mars. Analyses also show that a few rare meteorites contain material that is older than our solar system, and thus came from somewhere else—somewhere further out in space—perhaps from elsewhere in our Milky Way galaxy.

What's so exciting about meteorites is that most of them are very similar to the earth in composition. Even though they come from hundreds of millions of miles out in space, most of them contain minerals found on earth. And, the bulk of meteorites, called stony meteorites, contain generous amounts of silicon, one of the most common elements on the surface of the earth (the windows in just about every house in the world contain lots of silicon)! So, meteorites show us that the earth is made of "space materials."

Some meteorites bring with them little radioactive uranium clocks that were set when our solar system was born! The radioactive uranium found in these meteorites tells astronomers how old the meteorite is. Meteorite ages cluster around 4.5 billion years. This makes meteorites about 600 million years older than the oldest rocks found so far on earth (3.9 billion years).

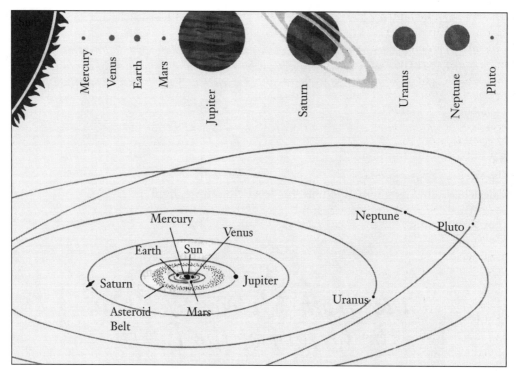

Figure II-4-1 The solar system showing the asteroid belt, the source of meteorites that hit the earth.
Source: F. Press and R. Siever, Earth, p. 5, Copyright 1978 W. H. Freeman and Co.

HERE ARE ANSWERS TO A FEW COMMON QUESTIONS:
You could share this information with your class, or you could assign the questions as research projects.

How many meteorites hit the earth? Every day an estimated 100 to 1,000 tons of meteorites hit the entire earth! (Forty tons is the average weight that a tractor-trailer truck can carry on today's highway.) We don't know how many meteorites fall, because most of them are very, very small. For every 100 to 1,000 of these meteorites, only *one* is large enough to hold between your fingers. So Grampa's meteorite is truly a treasure for Adam to protect!

How would you recognize a meteorite? With a magnet and a magnifying glass, you might be able to guess that a rock was a meteorite. Almost all meteorites are strongly to weakly magnetic depending upon their iron content. A few meteorites are strongly magnetic and extremely heavy when compared to terrestrial (from the earth) rocks of similar size. These are called iron meteorites and inside the iron minerals form an interlacing pattern that is not seen in earth rocks. The surface of some meteorites is pitted, resembling thumbprints in clay, and most meteorites have a streamlined shape because they were heated to nearly 3,000° F and the outer layer melted during their journey to earth.

Where do we find meteorites? Meteorites can be found anywhere. There is a statistically equal chance of them landing anywhere on earth, but they are easier to find where there are no trees. In 1969, Japanese scientists found large numbers of meteorites on the ice sheets in Antarctica, and since then thousands of them have been recovered and studied. Grampa suggested that their meteorite came to them as a gift from the glacier.

Are there meteorites in ancient rocks? Yes, meteorites have been bombarding the earth for billions of years.

Do all "shooting stars" end up as meteorites? No, in fact meteoroids that make it all the way through the earth's atmosphere and land on the surface are quite rare. Most "shooting stars" are comet fragments left behind in space like the con-trails of jet airplanes. The earth's orbit takes us through the comet debris at sometimes regular intervals of time, creating what are called "meteor showers."

IDEAS FOR ACTIVITIES:
- Call your state Geological Survey and find out what meteorites have been found in your state.
- Go to a local museum that has a meteorite.
- Make a meteorite out of papier-mâché. Cut it in half to show the inside.
- Find a website article about a meteorite and dinosaurs. What is the main point of the article?
- Hold class debate on what caused the great extinction of dinosaurs. Were there other mass extinctions in the earth's history? What caused them?

SOME QUESTIONS FOR FURTHER STUDENT RESEARCH:
We have given a bit of information on each, and more information is easily available from the Internet sites listed below:
- Who studies meteorites and why?
  Astronomers, physicists, and earth scientists all are interested in what meteorites can tell us about rock in outer space.
- What months do we see shooting stars? Why? Does the earth's orbit have anything to do with this?
  Shooting stars are common in the months of August and November.
- Does heat destroy magnetism?
  Yes, heat can destroy magnetism, but since many meteorites are magnetic, then apparently they haven't gotten hot enough.
- What human-made rock resembles a meteorite?
  Slag, the rock left in iron blast furnaces, resembles stony meteorites.
- What is iridium? Where is it found? Why is it important?
  Iridium is an element that some scientists use as evidence of meteors that have hit the earth in the past, because iridium usually comes only from outer space, or from volcanic eruptions.
- Where did the meteorite fall that may have killed dinosaurs? How big was it? Why do scientists think it killed them?
  The meteorite fell near Yucatan, Mexico.
- What does the surface of a meteorite look like?

REFERENCES AND FURTHER READING FOR TEACHERS/ADULTS:
Hansen, Michael, and Stig Bergstrom, 1997. ANCIENT METEORITES: Ohio Geology Quarterly Publication, Ohio Department of Natural Resources, Division of Geological Survey, Columbus, OH, 2 pp.
Norton, O.R., 1998. ROCKS FROM SPACE: Mountain Press Pub. Co., Missoula, MT, 449 pp.
Press, Frank and Raymond Siever, 1978. EARTH: W.H. Freeman and Company, New York, NY, 649 pp.
Renehan, Edward, Jr.,1996. SCIENCE ON THE WEB: A CONNOISSEUR'S GUIDE TO OVER 500 OF THE BEST, MOST USEFUL, AND MOST FUN SCIENCE WEBSITES: Springer-Verlag, New York, NY, 382 pp.

BOOKS FOR CHILDREN:
Cole, Joanna, 1987. THE MAGIC SCHOOL BUS, INSIDE THE EARTH: Scholastic, Inc., New York, NY, 40 pp.
McNulty, Faith, 1990. HOW TO DIG A HOLE TO THE OTHER SIDE OF THE WORLD: HarperCollins Publishers, New York, NY, 32 pp.

II-5

Birth and Early Childhood of the Earth and Our Fellow Planets

Because there are no rocks on earth as old as the solar system to tell us what happened, the story of the earth's beginning has been written and rewritten as new information comes to us from laboratories and from outer space. Planetary probes aboard American and Russian spacecraft have returned lots of information about the composition of Mercury, Venus, Mars, Jupiter, Saturn, Uranus, Neptune, and the Moon, and scientists have been startled to find that no two of our solar system planets are alike!

Here are some observations that need to be explained by any theory for the origin of our solar system of planets. 1) The planets all revolve around the sun in the same direction in nearly circular orbits that lie in nearly the same plane. Most of the moons also revolve around their planets in the same direction! 2) The planets, except for Venus and Uranus, rotate in the same direction as their revolution around the sun—that is, counterclockwise as one looks from the north pole to the south pole of the earth. 3) The distance of each planet from the sun is roughly twice that of the next planet closer to the sun (an ordering known as the Titius-Bode Rule).

One story that many scientists today believe explains these observations goes something like this: The early sun was surrounded by a disk of hot gas and dust that contained the ingredients for all the planets. As the disk cooled, more dust grains appeared and clumped together into planetesimals of various sizes. Some were large enough so that their gravity pulled the smaller ones into them, and they grew even bigger. In this fashion they all ended up rotating in the same direction and more or less in the same plane. Scientists think it took only about 100 million years for the solar system of planets to form. This is a very short time when you realize that the earth has been evolving for 4,500 million years!

REFERENCES AND FURTHER READING:
Press, Frank and Raymond Siever, 1978. EARTH: W.H. Freeman and Company, New York, NY, 649 pp.

USEFUL WEBSITES:

Moon
•http://asca.gsfc.nasa.gov/docs/StarChild/solar_system_level2/moon.html

Earth Science Educational Resources from Most U.S. States for K-6 Students
•http://www.kgs.ukans.edu/AASG/lists/elementary.html
The American Association of State Geologists provide sthis information on their website.

K-12 Science Lesson Plans Utilizing the World Wide Web
•http://www-sci.lib.uci.edu/SEP/CTS/

The Final Product
Our Water Planet with an Iron Core

OBJECTIVE: To show how the earth ended up with a dense, heavy, hot, iron-rich core material surrounded by the not-so-heavy, leftover mantle, and covered with continents, oceans, and the air we breathe.

The solid part of our earth is layered, and we live on a very thin outside layer called the crust. Underneath us, the layers get heavier toward the center and tell a story that goes back to the beginning of our solar system. Like the sonogram used to "see" a human baby in its mother's womb, sound waves are used by scientists to "see" inside the earth. Sound travels much faster through things that are hard. Try this: pound on a feather pillow, and you hardly hear any sound at all; most of it is absorbed into the spaces between the feathers. Now tap on a bell, and the sound comes back to you almost instantly. Scientists find that the earth is sort of like a large bell, and they listen for *big* sounds from things like earthquakes. With earphones—like the doctor's stethoscope—scientists all over the globe listen every time the sound waves from a huge earthquake go rumbling through the earth and out the other side. After many, many measurements they have learned that sound waves that go through the center of the earth travel much faster than waves that go near the center. The simplest explanation for this is that like a giant avocado, the earth has a solid core. (See Figure II-5-1.)

Other clever detective work shows that the earth is very heavy, and therefore that the center must be made of a heavy element. The most likely candidate is iron, since it is the most common heavy element in our solar system. But how can we have a solid iron core surrounded by a melted layer of iron in the center of our planet?

As you might suspect, the answer has to do with gravity, but there is a surprise in the story and it involves the moon. Scientists realized some time ago that the iron core of the earth had to have formed *before* the moon became our satellite, because without its iron core, the earth's gravity would not have maintained an attraction to the moon. It was a surprise when moon rocks brought back by Apollo astronauts told us that the moon was nearly as old as the earth. Researchers had to come up with a way to form the iron core fast. Geologically speaking, that meant within about 100 million years.

The early processes, theorized by geologists, were too slow to complete the iron core before the moon was formed. The early planet was a small, cold clump composed of a mixture resembling today's meteorites: about one-third iron and two-thirds lighter-weight rocky material. Each time a meteor crashed into the earth, it released some energy and squeezed the rest of the conglomeration tighter, which added heat. Small amounts of iron melted and started moving slowly toward the center. As the planet grew larger, the heat had further to go to get away, and more and more of it got trapped inside. At the same time, tiny radiocative

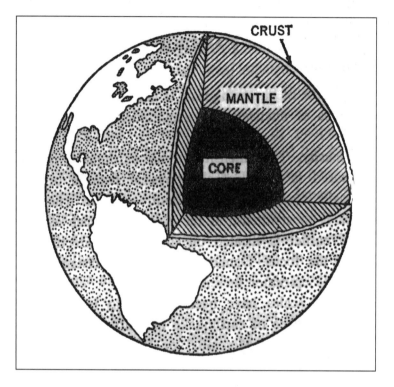

Figure II-5-1 Layers of the earth.
Source: W.H. Matthews, Geology Made Simple, *p. 104, Copyright 1967, Doubleday and Co.*

atoms added more heat by busily popping high energy particles out into the surrounding rock inside the growing planet. But it wasn't enough; researchers find that it would have taken too long—a billion years—before the young earth was hot enough to melt all the iron in our core!

To solve this dilemma, scientists made the interesting suggestion that a cosmic collision could have formed *both* the moon and the iron core at the same time! According to this idea, called "the giant impact," a very, very big cosmic body crashed into and melted much of the early earth. During the collision, material to make the moon was literally splashed out into space. Inside, gravity then took over, and the metallic and rocky parts may have separated like two liquids (for example, like oil and water) with the dense iron droplets raining through the molten rock and falling to the center of the earth, forming the core, while the lighter material floated upward toward the surface. The final result was an iron core and a surface crust made of lighter materials; between them was all the remaining material in what is called the mantle.

The crust of the earth—the part we walk on—is light (one-half of it is made of oxygen!), and it surrounds the mantle sort of like an orange peel. Scientists think that since the earth formed, the crust has been getting lighter and lighter as the heavy material sinks to the center of the earth. Like a furnace, the earth's internal heat engine drives this process, which continues today. The continents float on the heavier mantle like huge ships.

Overhead, our atmosphere of oxygen and nitrogen was a gift from zillions of tiny organisms called cyanobacteria, or blue-green algae. Imagine getting the air we breathe from a thing so small you can only see it with a microscope! Scientists think that at first the earth had no atmosphere because the planetesimals didn't have any. So gas must have come from inside the earth, which was "burping" like a shaken-up Coca-Cola. This original atmosphere resembled the gases that scientists in fireproof space suits have collected from volcanoes—water vapor, hydrogen, hydrogen chloride, carbon monoxide, carbon dioxide, and nitrogen. We couldn't have survived in this early "air" without a gas mask! Little tiny organisms used the carbon dioxide from this early atmosphere, along with the energy of the sun, and made oxygen. Over the millennia since the cyanobacteria started their work, our atmosphere has been changing into the air that blows our willow trees and keeps airplanes flying high.

And finally, the oceans of our blue planet—where did they come from? To answer this question, we need to look for the source of the water in our oceans. It came as a surprise to recently learn that water is common out in the universe. Some scientists therefore suggest that the water on earth may have come from comets. These heavenly bodies are composed largely of frozen water and carbon dioxide. Other researchers feel that our oceans gradually accumulated from volcanic eruptions early in the earth's history. These scientists observed abundant water in modern lava and have calculated that there is enough to form the oceans. Like some gigantic steam generator, water boiled up into the atmosphere and crust from minerals in the cosmic dust of the early earth. Water was locked up inside minerals like mica $[KAl_3Si_3O_{10}(OH)_2]$ and was freed when the mica melted into magma inside the earth. Volcanoes brought the water to the surface with lava.

ACTIVITY I

Scale Model I of the Earth-Moon System

OBJECTIVES:
1. To show students the relationship between the earth and its moon with respect to size and distance.
2. To give students opportunities to make a model and apply math skills.

> National Science Education Content Standard(s), K-4, 5-8
> Major Emphasis: A, C, E; Minor Emphasis: B;
> Other Content: Mathematics, Scale

INFORMATION NEEDED:
Diameter of the earth = 12,800 km
Diameter of the moon = 3,500 km
Mean distance from earth to moon = 385,000 km

MATERIALS NEEDED:
Masking tape; a long piece of heavy string; a basketball; a tennis ball

PROCEDURE:
(Answers for the teachers are provided in brackets.)
1. Calculate to the nearest whole number the number of moon diameters it would take to equal the diameter of the earth. [Answer: For any scale model, the earth must be four times as large as the moon.] If we choose a basketball (diameter of approximately 25 cm) to represent the earth, what size ball would represent the moon? [Answer: approximately 6 cm, a tennis ball.]
2. Calculate how many earth diameters (12,800 km) equal the distance to the moon (385,000 km)? [Answer: Approximately 30 earth diameters.] If one earth diameter is represented by a basketball, how many basketballs must fit between the earth and the moon? [Answer: For any scale model, the distance between the earth and the moon must be 30 "earths."]
3. Now obtain a long piece of heavy string. Using our scale, how long should the string be to represent the distance between the earth and the moon? [Answer: 30 x 25 cm = 750 cm = 7.5 meters.]
4. With masking tape, attach one end of the string to the "earth" (basketball) and the other end to the

"moon" (tennis ball). You now have a very flexible earth-moon model which is to scale for both size and distance—the earth and moon in your classroom!

5. With this model, how far away would the sun be, in units of our earth-moon distance? [Answer: 400.] Hint: the earth-sun distance is 150,000,000 km. If we took our model outside into the street, how far away would the sun be in meters? [Answer: 400 x 7.5 meters = 3,000 meters = 3 km.] This exercise should provide the students some food for thought as to the immense size and vast emptiness of our solar system!

ACTIVITY II

Scale Model II of the Earth-Moon System

OBJECTIVE: To demonstrate lunar and solar eclipses.

> National Science Education Content Standard(s), K-4, 5-8
> Major Emphasis: A, C, E; Minor Emphasis: B;
> Other Content: Mathematics, Scale

MATERIALS NEEDED:
Two Styrofoam balls, one four times larger in diameter; wood; rubber bands; paperclips; glue.

PROCEDURE:
(Answers for the teachers are provided in brackets.)

1. Let the earth be represented by a 10 cm styrofoam ball. What size styrofoam ball should represent the moon? [Answer: 2.5 cm. Our scaling factor for size is 4.]

2. What is the earth-to-moon distance with our new scale? [Answer: 30 x 10 cm = 300 cm = 3 meters. Remember, our scaling factor for distance is 30.] Go to your local lumberyard and have them cut an inexpensive 1-inch x 1-inch board whose length correctly represents this earth-moon distance.

3. Attach the larger ball (the earth) to one end of the board with glue or large rubber bands. Place the smaller ball (the moon) at the other end. In order to keep the shadow of the moon distinct from the shadow of the board, raise the smaller ball above the surface of the board by attaching it to a nail or a stiff wire. An unbent paper clip works well.

4. Armed with this model (and it is an armful), take your class outside on a sunny day and demonstrate eclipses. It really works! You will discover that the "moon" will be completely covered by the "earth's" umbra as you simulate a lunar eclipse just as in the actual event, but during your "solar" eclipse only a portion of the "earth's" surface will fall into the "moon's" total shadow. Again, this is what actually happens. Our model illustrates nicely why one must live within that narrow zone of totality in order to experience a total solar eclipse. The strip of totality is generally only about 241 km wide, which explains why most people have never seen a total solar eclipse.

REFERENCES AND FURTHER READING:
Cole, Joanna, 1987. THE MAGIC SCHOOL BUS, INSIDE THE EARTH: Scholastic, Inc., New York, NY, 40 pp.
Hansgen, Richard, 1996. TWO SCALE MODELS OF THE EARTH-MOON SYSTEM: Hands On Earth Science #10, Ohio Department of Natural Resources, Division of Geological Survey, Columbus, OH.
Matthews, William III, 1967. GEOLOGY MADE SIMPLE: Doubleday & Company, Inc. New York, NY, 192 pp.
McNulty, Faith, 1990. HOW TO DIG A HOLE TO THE OTHER SIDE OF THE WORLD: HarperCollins Publishers, New York, NY, 32 pp.
Press, Frank and Raymond Siever, 1998. UNDERSTANDING EARTH: W.H. Freeman and Company, New York, 682 pp.

K-12 Teaching About the Earth
•http://www.athena.ivv.nasa.gov
NASA's Athena site for teachers and students, funded by public use of remote sensing data, features oceans, earth resources (earthquakes, landforms, wetlands, global change, Landsat images), weather, and space and astronomy.

Moon
•http://asca.gsfc.nasa.gov/docs/StarChild/solar_system_level2/moon.html

The Jason Project
•http://www.jasonproject.org/front.html
Explore the oceans with Bob Ballard.

OUR DYNAMIC EARTH

III-1

The Story of the Rock Cycle

OBJECTIVE: To help students discover that dynamic processes have shaped the earth for billions of years. Under the force of gravity, rocks on the surface of the earth are constantly in motion and are constantly changing from larger to smaller pieces and back to larger pieces again. The smallest grain of sand on a mountain top yearns to go down to the ocean, while centimeter by centimeter the mountains are being pushed upward by enormous pressures. Inside the mountains, sandy ocean beaches are being tortured into rock-hard quartzite. (See the Rock Cycle in Figure III-1-1.)

> *"Look closely," said Grampa, offering the stone to the boy. "See what's trapped inside."*
>
> *Adam turned the grainy stone in his hands like a kaleidoscope, squinting from the reflected sunlight. He glanced up at Grampa, then down at the stone, then back up again, not knowing what to say next. Finally, he saw what had been there all along, grains of clean sand, now frozen within the stone.*
>
> *"Looks like sand to me," said the boy quietly, still unsure of himself.*
>
> *"Right!" confirmed Grampa. "It's beach sand. From an ancient ocean that was here, right here, a very long time ago."*

From Stone Wall Secrets

The sparkling, white sandstone that caught Adam's eye was made of quartz grains, and that's where our story begins. It is a story about how the rock in Grampa's wall was once an ancient quartz sand beach along an ocean that no longer exists. Quartz is a mineral that is very hard for the forces of nature to break down. Like much of the dust in space, quartz is made of the same elements as window glass (SiO_2, or silicon dioxide, which means it is made of one atom of silicon for every two atoms of oxygen). A common source of quartz is the rock, granite, (shown in Figure III-1-2) in which the quartz is joined by two other minerals, feldspar and mica (or often, the dark mineral hornblende). When granite breaks apart, sparkling, light gray quartz is often the only mineral to survive a long, bumpy journey to the beach. The feldspar and mica break apart and end up in tiny bits as soil along river banks and mud that washes way out to sea. Because the quartz stays in slightly larger pieces (called sand grains, see Size Table), it is harder for ocean currents to move it away from the land. That is why ocean beaches are often made entirely of pure quartz sand.

But that's only the beginning of our story. Now that the quartz has arrived at the beach, how does the sandy beach become a sandstone? And how did a piece of sandstone from this ancient beach find its way into Adam and Grampa's stone wall?

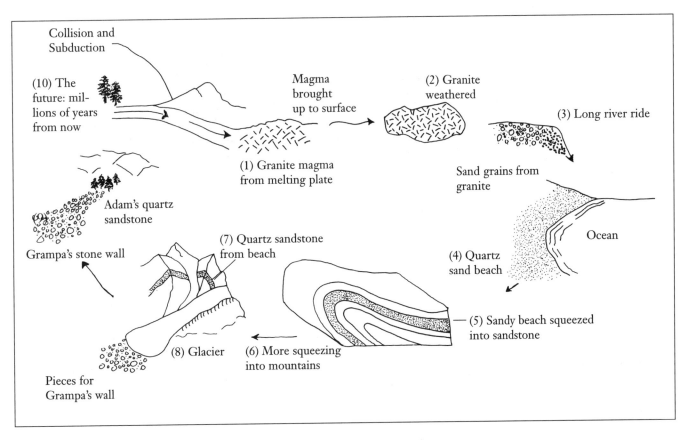

Figure III-1-1 The Rock Cycle

Rock Cycle showing granite to sandy beach to sandstone. Then mountain-building, glaciation, and finally Grampa's stone wall. This illustration will help younger students (like Adam) see the connection between stone walls and ancient beaches.

From (1) to (10) in several hundred million years!

> *(1) Granite intruded beneath continent (eg. from a subducting plate)*
> *(2) Granite (quartz, feldspar & dark minerals) is exposed to weathering*
> *(3) Long river ride—feldspar and dark minerals turn into mud, quartz becomes sand*
> *(4) Mud goes out to sea, quartz sand stays near shore to make beach*
> *(5) Sandy beach buried and then squeezed into sandstone and or quartzite by tectonic collision along coast*
> *(6) More squeezing makes really big mountains*
> *(7) Sandstone layer exposed on mountain side and eroded by glacier*
> *(8) Glacier makes slabs for stone walls*
> *(9) Glacier dumps rocks, frost heave brings them up, and farmers use them to make stone walls*
> *(10) Stone walls are eroded and turned into granite and the cycle starts all over*

Figure III-2 Grains of quartz weather out of the granite to make the sandy beach and then the sandstone.
Source: M. Bramwell, Rocks and Fossils, *p. 8 and 11, Copyright 1994, 1983, Usborne Publishing, Ltd.*

	Size Table	
	How Big is a Cobble?	
	Table showing young students how to recognize approximate sediment sizes.	
Name	*How to Determine Size*	*Approximate Size, Inches*
Boulder	Takes both hands to lift the smallest ones	Greater than 10
Cobble	Can hold most in one hand	Greater than $2^1/_2$ to 10
Pebble	Can hold between thumb and finger	Greater than $^1/_{10}$ to $2^1/_2$
Sand Grains	Feels gritty between finger and thumb	
Clay	Can't see grains — Doesn't grit between teeth	

"Eons passed," he went on, now pointing at the opposite ridge. "The mud and sand were buried deep within the earth. There, miles below the surface, heat and pressure, bit by bit, baked the mud into slabby, gray rock—rock that now fills all these walls."

From *Stone Wall Secrets*

To change from beach sand to sandstone, the space between the sand grains must be filled with one of nature's cementing glues. The cementing of sandstone starts to happen after the beach has been buried, and there are tons of sediment pressing it down. (See Figure III-1-3.) Sometimes, if the pressure gets great enough, the quartz grains begin to actually melt into each other. This rock is called a "quartzite." Pretty soon the only clue to the gleaming shores of the ancient ocean is a layer of very hard rock made of sparkling quartz.

The journey from deep in the earth to Grampa's stone wall is far more perilous than a river ride down to the ocean. Humongous forces are needed to lift the sandstone layer back up into the sunlight. To bring Adam his rock required a continental collision! (I can hear the question: What's a continental collision? We'll talk about this soon, but for now look at a globe of the earth and imagine two continents bumping into each other.) While the sand grains from Grampa's ancient beach were deep within the earth, the New England coast of North America was repeatedly crushed by the slow, westward movement of a huge continental mass. Now, when one huge continent collides with another continent something has to give, and the layers of rock containing our beach were buckled upward into mountains (see Rock Cycle, Stages 5 and 6, Figure III-1-1.) that rivaled today's Rocky Mountains. These early New England mountains, however, are about 450 million years old!

The sparkling white sandstone has been lifted up, but it is now trapped inside a huge mountain range. How in the world did it get to Grampa and Adam's wall? As Grampa tells Adam, it took a very long, long time. Long enough, in fact, for the mountains to fall apart, grain by grain!

"But mountains come and mountains go. Making them is the easy part; getting rid of them is much harder. Nature needs time, lots of time, for the sun, the wind, and the rain to break the rocks apart and send them to the sea."

From *Stone Wall Secrets*

Over the next hundreds of millions of years the mountains were worn down by freezing winters, spring winds, and summer thunderstorms, and eventually just the hardest layers were left beneath the rolling hills of New England. Among these hard layers was the sandstone—the once loose, sandy beach that now would become Adam's white, sparkling rock. It would take a glacier, Mother Nature's "bulldozer," to wrestle with the hard sandstone and bring Adam and Grampa their rock.

The forces of nature never rest, and now the very old mountain rocks were broken into blocks by glaciers that relentlessly ground across the countryside. (See Rock Cycle Figure III-1-1.) When the glaciers melted, rivers of muddy water carried pieces of Adam's sandstone to all parts of the New England countryside. The blocky rocks, along with a jumble of mud and sand, were unceremoniously dumped into the area that would become Grampa's field. Thousands of years later it was one of many that made their appearance in the muddy New England springtime and ended up as part of Grampa's stone wall.

Millions and millions of years from today, Grampa's wall may be buried yet again under layers of new sediment, and in millions of years more, it may rise up again to be eroded by yet another glacier. This is called a *cycle*, something that goes around and around, and this cycle is called the Rock Cycle. The Scotsman

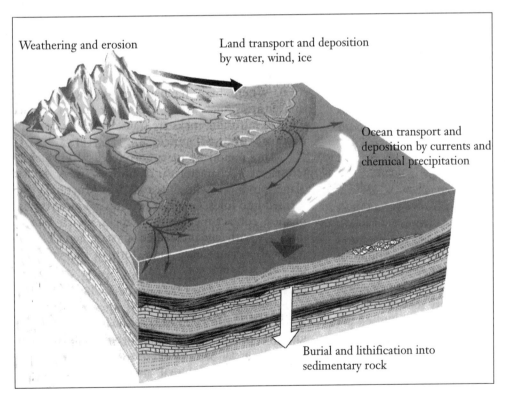

Figure III-1-3. Weathering breaks granite down into minerals. Quartz generally survives the journey to the beach. Sediment piles up for millions of years, and when the beach is deeply buried, it starts to turn to sandstone.
Source: F. Press & R. Siever, Understanding Earth, *Fig 3.3, p. 63, Copyright 1998, W. H. Freeman & Co.*

James Hutton described it in an oral presentation in 1785 before the Royal Society in Edinburgh; ten years later he presented it in more detail in his book, *Theory of the Earth with Proof and Illustrations*. The story of how he learned about the Rock Cycle is a fascinating one.

ACTIVITY I

Class Discussion on the Origin of Sand

OBJECTIVE: To provide the opportunity to focus the whole class on an important earth process: the formation of sand. This will help students to understand the energy needed for mechanical erosion and the source material for the sand.

> National Science Education Content Standard(s), K-4, 5-8
> Major Emphasis: A, C; Minor Emphasis: B, D, G, H;
> Other Content: Public Speaking, Geography

MATERIALS NEEDED: Five hand samples of sandstone; globe or world map showing climate areas (desert, tropics, Arctic, etc.); encyclopedia; rock and mineral reference book with definitions (see Bramwell, *1994 Usborne Guide*, complete reference below).

PROCEDURE:

What is sandstone? Divide the class into groups of no more than five or six students. Give each group a sample of sandstone and ask them to describe five things about it. Give groups fifteen minutes, then have student from each group give their findings to the class and list them on the board.

Where does sand come from? Have groups come up with three ways Nature can "make" sand; e.g., an ocean beach, sand dunes, etc. Give groups fifteen minutes, then have students from each group give their findings to the class and list them on the board. Identify the energy source in each case (wind, water, glacier, etc.).

Where would you go to find sand? Put this question on the board for the whole class to answer. Use a globe or a world map to identify places on the present earth most likely to have accumulations of sand. Try to pick the places based on class criteria. Divide back into the original groups and have each group check class results with an encyclopedia, or other source of information. The goal is to have the class get information on the specific locations identified on the basis of your class criteria for sand-making regions.

ACTIVITY II

Hands-On Examination of Sandstone and Quartzite

OBJECTIVE: To determine the difference between these two rock types, and to demonstrate that it isn't always easy to tell them apart!

> National Science Education Content Standard(s), K-4, 5-8
> Major Emphasis: B, C, E; Minor Emphasis: A, H;
> Other Content: Public Speaking

MATERIALS NEEDED: Five hand samples of sandstone and five samples of quartzite; at least three rock and mineral reference books with definitions of rocks. (See Bramwell, *Rocks and Fossils*, complete reference below.)

PROCEDURE:

At what point does a rock cease being a sandstone and become quartzite? [Answer: It is hard to tell, and one must decide which rock definition to use.]

1. Divide the class into groups of no more than five or six students. Give each group a magnifying glass

and samples of sandstone and quartzite. Ask them to describe five things about each. Give groups ten minutes, then have a student from each group give the group's findings to the class and list them on the board.

2. With the class as a whole, get several definitions for sandstone and quartzite. Use as many sources as you can. List on the board the things common to each definition. There are only a couple—keep it as simple as you can.

4. Now have the groups re-form and see if they still agree that their rocks are sandstone or quartzite.

ACTIVITY III

Reliving James Hutton's Talk Before the Royal Society in Edinburgh, Scotland

OBJECTIVE: To learn about the history of science and to show that Hutton introduced the concept of geologic time.

> National Science Education Content Standard(s), K-4, 5-8
> Major Emphasis: A, C, E, H; Minor Emphasis: B;
> Other Content: Public Speaking

PROCEDURE:
Have your class present quotes from Hutton's talk, *Theory of the Earth with Proof and Illustrations*. Dress up in Period costumes. This is a great way to study history *and* the Rock Cycle. Two famous quotes are:

"Time, which measures every thing is our idea and often deficient to our schemes, is to nature endless and as nothing." (Hutton, 1788, p. 215.)

"If the succession of worlds is established in the system of nature, it is in vain to look for anything higher than the origin of the earth. The result, therefore, of our present inquiry is that we find no vestige of a beginning—no prospect of an end." (Hutton, 1788, p. 304.)

REFERENCES AND FURTHER READING:
Hutton, J., 1788. THEORY OF THE EARTH. Transactions of the Royal Society of Edinburgh 1: p. 209-305.
More information about James Hutton and how he discovered the rock cycle may be found in the following reference:
Gould, Stephen Jay, 1988. TIME'S ARROW/TIME'S CYCLE: MYTH AND METAPHOR IN THE DISCOVERY OF GEOLOGICAL TIME (Jerusalem-Harvard Lectures); Harvard University Press.

ACTIVITY IV

A Fun Exercise for Learning to Identify Rocks

OBJECTIVE: To learn to identify rocks.

PROCEDURE:
Divide your class in half. Each group hides ten things they would like to find in a stone wall. Then the other half of the class searches for them. The catch is that the ten things must have something to do with earth science. That is the only rule; the "things" can be drawings, pictures, rocks, minerals, fossils, etc.

DISCUSSION:
It is important to keep this activity fun; leave plenty of time for questions.

ACTIVITY V

Reading the Rocks—They Can Tell Us About the Earth Millions of Years Ago

OBJECTIVE: To learn to identify the environments in which different sedimentary rocks were formed.

> National Science Education Content Standard(s), K-4, 5-8
> Major Emphasis: A, C, E, G; Minor Emphasis: B, H;
> Other Content: Public Speaking

MATERIALS NEEDED:
- A generous collection of sedimentary rocks with known depositional environments
- Writing tablets for notes (stenographers notepads work well)

PROCEDURE:
1. Place a rock and a notepad on each student's desk and write on the blackboard the following question: "Where do you think the rock on your desk was formed?" Have the students write what they think in the notepad. Be sure each student puts his or her name on his or her ideas. (The notepad will stay on the desk with the rock.)
2. After about five minutes, let the students get out of their seats and examine the rocks on each of the other desks. Have them write their ideas in each of the notepads on each desk.
3. When as many students as possible have had a chance to look at the rocks, have them return to their desks. Now have each student in turn "Read the Rocks." Have him (or her) bring the notepad and rock from his desk up to the front of the room and tell what he thinks about his sedimentary rock. Then have him read what the other students have said.

Following are some examples of sedimentary rocks and the environments in which they formed:
- *Black shale* with leaf impressions [ancient swamp]
- *Breccia*, a rock with angular pieces in hardened mud [ancient landslide]
- Pebble conglomerate, a mud or sand rock (mudstone or sandstone) with rounded pebbles in it [a big river or a glacier]
- *Fossil limestone*, a lime rock (it will fizz with acid) that has fossils in it [an ancient ocean, probably in a warm climate]
- *Halit*e, a rock made of salt [a place where salty lake or ocean water was evaporating, hot, dry, like Great Salt Lake in Utah, or the Dead Sea in Israel]
- *Shale*, a rock made of mud with the water squeezed out [way out on the ocean floor where big rivers flow into the ocean, like where the Mississippi flows into the Gulf of Mexico]
- *Sandstone*, a rock made of sand grains [an ancient beach; a sand dune]
- *Shale with fish fossils*, a light-colored, soft rock in thin layers that show rusty fish fossils [from a huge ancient lake in Wyoming—many layers of shale formed over millions of years, each layer tells of a time when most of the fish in the lake died off suddenly and were buried beneath the next layer of mud]. This is a well-known rock shop sample.

REFERENCES AND FURTHER READING:

Bramwell, Martin, 1994. Rocks and Fossils: EDC Publishing Ltd., London, UK, 31 pp.

Powell, John Wesley, 1978. EXPLORATION OF THE COLORADO RIVER: U.S. Dept. of the Interior/U.S. Geological Survey, U.S. Government Printing Office, Washington, DC, 28 pp.

Rock Detective, Inc., 1998. ROCK MYSTERIES PUBLICATION: Rock Detective, Inc., Dresden Mills, ME 04342, 165 pp.

The Society for Sedimentary Geology offers K-12 educators the following excellent publications which can be ordered by calling 800-865-9765.

Miller, Molly F. , R. Heather Macdonald, Linda E. Okland, Steven R. Roof, and Lauret E. Savoy, 1990.

SEDIMENTARY GEOLOGIST'S GUIDE TO HELPING K-12 EARTH SCIENCE TEACHERS: HINTS, IDEAS, ACTIVITIES AND RESOURCES, 92 pp., paperbound, ISBN 0-918985-86-2. Catalog #81001. Price $5.00.

Macdonald, R. Heather and S.G. Stover, 1991. HANDS-ON GEOLOGY: K-12 ACTIVITIES AND RESOURCES, 105 pp., spiralbound, ISBN 0-918985-90-0. Catalog #81002. Price $6.00.

Stover, S. G. , and R. Heather Macdonald, 1993. ON THE ROCKS: EARTH SCIENCE ACTIVITIES FOR GRADES 1-8: 204 pp., spiralbound, ISBN 1-56576-005-0, Catalog #81003. Price $9.00.

USEFUL WEBSITES:

Kids Web for Earth Science
•http://www.ucmp.berkeley.edu/fosrec/TimeScale.html
"Front door" for student-oriented current events, includes, Ask a Geologist, earthquake information, current research.

The Rock Cycle
•http://www.science.ubc.ca/~geol202/rock_cycle/rockcycle.html

Resources for Teaching About Sedimentary Rocks
•http://www.beloit.edu/~SEPM/
A website dedicated to K-12 educators, geoscientists involved in K-12 education, and students interested in careers in sedimentary geology.

Scientific Resources Online
•http://www.scicentral.com
SciCentral is maintained by professional scientists whose mission is to identify and centralize access to valuable scientific resources online. Currently the page has more than 50,000 sites pertaining to more than 120 specialties in science and engineering, including science in the news and a K-12 section for educators.

Earth Science Information
•http://www.gsfc.nasa.gov
NASA, Mission to Planet Earth.

Rocks, Fossils, and Minerals for Teaching Earth Science
•http://www.rmid.com/rocks
A non-profit organization of teachers and earth scientists, Rock Detective, Inc., offers inexpensive teaching samples. In addition, as described on the website, Rock Detective offers a variety of teacher-friendly, hands-on earth science programs customized for K-12 students.

III-2
Why the Sea is Salty

OBJECTIVE: To introduce two concepts:
1. That elements make minerals which in turn make rocks; and
2. The dynamic earth acts to break the minerals apart and free the elements to go to the ocean.

Why is the sea salty? The answer to this question is hiding in the answer to another question: Did you ever wonder what happens to the sugar crystals in a cup of tea? The crystals disappear, but where do they go? Well, the water pries the sugar apart, into elements like carbon and hydrogen. This very same thing happens to the minerals in mountain streams and big rivers. Like lumps of sugar, the minerals are made of grains. Water flowing down the stream pries the mineral grains apart, and like the sugar in our tea, the *elements* that make up the minerals disappear into the water, or *dissolve*. By the way, water is like Mother Nature's packrat; it loves to carry away dissolved elements. After a while, the dissolved mineral tidbits work their way into bigger and bigger rivers, and finally down to the ocean. Of course, there is very little sugar in the rocks and soils around a stream, so the ocean is salty, not sweet.

So what is it about sugar and minerals that makes it possible for water to pry them apart? And what's a mineral anyway? We need to start by looking at the elements that minerals are built from. An element is like one of the stones in Grampa's wall; it is the building block of matter. The elements are locked together inside each mineral grain, the way Grampa and then Adam learned to pack the stones in their wall so they wouldn't come apart. Remember H₂O? That's the way we describe the elements inside water: two atoms of hydrogen and one atom of oxygen.

Because the ocean is salty, let's first look at the elements inside salt.

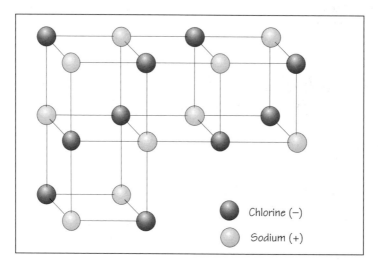

Chlorine (−)

Sodium (+)

Figure III-2-1. Atoms of the elements chlorine (Cl) and sodium (Na) inside the mineral halite, or table salt.
Source: Dr. Tom Freeman, Friendship Publications, Columbia, Missouri. Copyright 1992. Used by permission.

Salt, just like table salt, is a real mineral. Its name is halite, and it has two elements locked together inside. The two elements are called sodium (Na) and chlorine (Cl). When there isn't any water around, the two elements hold tightly to each other and we see white grains of salt. But when water enters the picture, it grabs hold of the elements and carries them away. Now, like a convertible car, the water carries the sodium and chlorine around in the back seat. That's right, the sodium and chlorine are riding around in the back seat of the water *molecules*.

Gradually, over millions and millions of years, elements riding in the back seats of zillions of water molecules end up in the ocean. The elements all come from minerals, and the minerals all come from rocks, like Adam's white sparkling sandstone, or the gray rock from Grampa's wall, made from layer upon layer of mud.

ACTIVITY I

Where Does Our Table Salt Come From?

OBJECTIVE: To begin to understand that layers of halite formed from evaporated sea water on the surface of the earth get buried and millions of years later mined for salt.

> National Science Education Content Standard(s), K-4, 5-8
> Major Emphasis: A, E; Minor Emphasis: B, C, G;
> Other Content: Geography, Public Speaking

MATERIALS NEEDED: A globe; several references; Internet

PROCEDURE:
1. Divide class into several groups and have each group use a reference (e.g., Internet, encyclopedia, geographic atlas) to find at least one place in the world where salt is mined. (Have advanced students try to determine the geological age of the halite.)
2. Have each group give its results and, at the same time, locate each place on a globe large enough for the class to see.
3. In each case see if the class can figure out the source of the salt. Does the globe offer any clues about where the salt came from?

ACTIVITY II

Is it a Rock or a Mineral? And Who Decides What Things Will Be Called?

OBJECTIVE: To demonstrate the difference between rocks and minerals *and* between *definitions* of rocks and minerals; to create a definition.

> National Science Education Content Standard(s), K-4, 5-8
> Major Emphasis: A, B, H; Minor Emphasis: E;
> Other Content: Language Arts, Public Speaking

Note: This important activity takes students out of the traditional mode of having to just learn and then remember definitions. It gives them a chance to create their own definition—and then to defend it!

MATERIALS NEEDED:
"Minerals": Hershey Kisses™, gummy bears, jelly beans, chocolate &/or peanut butter chips.
"Rocks": Peanut M& M's™, Nestle's Buncha Crunch™, Butterfinger BB's™, Hershey Kisses with Almonds™
(This activity assumes that none of the students are diabetic or allergic to chocolate, peanuts, or almonds.)

PROCEDURE:
1. Discuss the definitions of rocks and minerals. Obtain several definitions for rocks and several definitions for minerals [see the references listed below].
   Some Reference Ideas for Finding Rock and Mineral Definitions
   • Various dictionaries
   • Glossary, Stone Wall Secrets Teacher's Guide
   • Wooley, A., Usborne Spotter's Guide—Rocks and Minerals: Usborne, ISBN (P) 0-86020-112-0, 64 pp.
   • Bramwell, M., An Usborne Guide—Rocks and Fossils: Usborne, ISBN (P) 0-7460-1975-0, 32 pp.
   • Bates, R.L. and J.A. Jackson, Eds., 1987. GLOSSARY OF GEOLOGY: American Geological Institute, Alexandria, VA, 788 pp.

2. Write the main concepts from each definition on the board—have the students copy them as you write. Look for inconsistencies between the definitions. Now have your students create their own definition that incorporates the consistent part of the definitions you have just looked at. They will have to make some decisions, for example, whether oil is a mineral, or whether concrete is a rock. Now have your students name their definitions! It is very important that they name the definitions. This will help them realize that they are making their own choices about the definition that they choose to use, and that they can change at any time and use another definition just as long as they give each definition a name and tell which one they are using (and why).

3. Now, divide your class into groups and pass out the candy—make sure each group gets several of each type. Have the students write a description of each candy sample and then "classify" the candies as either "rock-like" or "mineral-like" according to their definitions. They will need to see the inside of their "sample," so they will have to take a bite (but don't eat it all!).

4. When all the groups have finished, have a spokesperson from each group name the candy samples as either "rock-like" or "mineral-like" and tell why. Save time for this because there will be plenty of debate. Make sure the definitions created earlier are used.

5. Last, have each student bring in a sample of a rock and one of a mineral. They can bring them from home, or you can take the class outdoors and find samples from the schoolyard. Have each student tell the class why he or she thinks the sample is a rock or mineral. Put those that can't be identified in a category called "unknown." Maybe later someone will figure out what they are. You will probably find that you have many more rocks than minerals, simply because there are more of the former in nature. If you don't have enough minerals, it would be a good idea to have your school invest in a good rock and mineral collection. You can hide them around the classroom and have a treasure hunt! Then talk about what they are.

ACTIVITY III

Crystal Garden

SOURCE: Weisgarber, Sherry L., *Crystal Garden, Hands On Earth Science, #10*, Ohio Department of Natural Resources, Division of Geological Survey, Columbus, Ohio. Taken from: *Kids Create!*, Laurie Carlson; and Nevada Mining Association, Lois K. Ports

OBJECTIVE: To show that crystals, like mineral grains, can come *out of* water.

National Science Education Content Standard(s), K-4, 5-8
Major Emphasis: C; Minor Emphasis: B

MATERIALS NEEDED:
6-7 untreated barbecue charcoal briquettes, or stones (1 to 2 inches across)
Shallow bowl
4-6 tablespoons table salt
4-6 tablespoons liquid laundry bluing (see note below)
4-6 tablespoons water
1 tablespoon ammonia (be careful using ammonia around children)
Food coloring

PROCEDURE:
Collect several small pieces of limestone, brick, coal, or barbecue charcoal. You may want to try a bowl of each to determine which material grows the best crystals. Place the charcoal or stones clustered together in the bowl. Mix all of the ingredients together, except the food coloring, in the order listed using the same amount of salt, bluing, and water for each batch made. Pour the mixture very slowly over the stones with a spoon. The mixture may not be dissolved depending on the number of tablespoons of ingredients used. You may want to make different batches using different amounts of ingredients to see which works best. Drop food coloring over the coated stones. Using different colors produces a variegated crystal garden. Crystals should begin to form in about twenty minutes and continue growing for a day or two. Adding any excess mixture to the bottom of the bowl over the next few days may keep the garden growing longer. This creation will crumble very easily, so don't move it around too much.

Note: Laundry bluing comes in a small blue bottle and generally can be found in the laundry section of a grocery store next to the starch and bleach products.

REFERENCES AND FURTHER READING:

Kendall, David, 1993. GLACIERS & GRANITE, A GUIDE TO MAINE'S LANDSCAPE & GEOLOGY: North County Press, Unity, ME, 240 pp.

Press, Frank and Siever, Raymond, 1998. UNDERSTANDING EARTH: W.H. Freeman and Company, New York, NY, 682 pp.

USEFUL WEBSITES:

Teacher Workshop/Field Experience
•http://cgee.hamline.edu/rivers
Rivers of Life: Mississippi Adventure, produced by Hamline University's Center for Global Environmental Education, offers Rivers of Life, an exciting hands-on project that gets students outside to experience the drama of spring while learning about where our water comes from and where it goes. You can preview the ROL classroom at the website above.

U.S. Geological Survey Learning Web
•http://www.usgs.gov/education/
A wide variety of useful material, activities, and resources for teaching earth science.

Lesson Plans
•http://atm.geo.nsf.gov/
Lesson plans and classroom materials utilizing real-time environmental information in earth science and mathematics classes.

III-3

Munching Plates and Crunching Continents

OBJECTIVE: To show that throughout geologic time continents and the tectonic plates they ride upon have been pushed about and have collided relentlessly with each other. Collisions between huge pieces of the crust of the earth are still happening, just as they have for billions of years — and, with the same frightening and beautiful results: devastating earthquakes, hundreds of active volcanoes, and lofty, snow-covered mountains.

Now this is a story that will knock your socks off! Did you know that next year if you were to come back to this very spot, it would have moved a few inches, or centimeters, in one direction or another? That's right! Satellites can now measure very, very accurately just where things are on the earth, and scientists have learned that most of the continents are moving quite quickly, geologically speaking.

So how can something as large as a continent *move?* Well, the whole surface of the earth is broken into huge pieces called tectonic plates that are munching and crunching into each other. The continents are riding along on these plates, and sometimes they move away from each other, and sometimes they bump into each other. The tectonic plates and their movements are shown in Figure III-3-1.

On a globe of the earth we can easily find two continents that have moved apart. First, find South America. Now find Africa. See how they look like pieces of a gigantic jigsaw puzzle? (Activity I will help you to see this.) At one time these two huge continents fit together and were part of one land mass called Pangaea (See Illustration III-3-2). South America and Africa have moved apart (See Illustration III-3-3)— and they are still moving! Today the Atlantic Ocean is getting larger.

Plates on the earth have been moving since the crust and mantle were formed. In the past, millions of years ago, long before Pangaea, the continents were different shapes and were located in different parts of the earth. Ancient oceans where the Atlantic is today, east of the United States, slowly *closed* and are gone. The ocean floor literally slid underneath North America! Continents riding on these ocean plates collided with North America and changed the shape of our land.

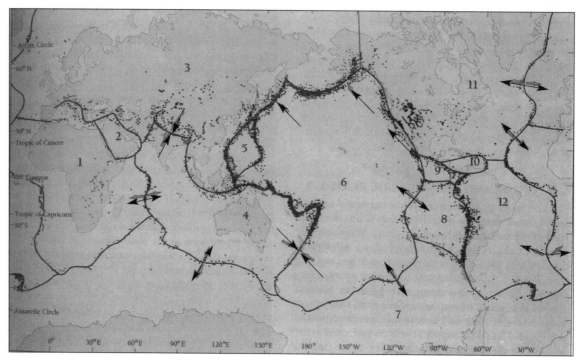

1. African plate	4. Australian plate	7. Antarctic plate	10. Caribbean plate
2. Arabian plate	5. Philippine plate	8. Nazca plate	11. North American plate
3. Eurasian plate	6. Pacific plate	9. Cocos plate	12. South American plate

Figure III-3-1. Tectonic Plates of the Globe. Arrows show plate movement.
Source: P. R. Pinet, Oceanography, *p. 73. 1992. Adapted by Pinet, from K. S. Stowe,* Ocean Science
(New York: John Wiley, 1983).

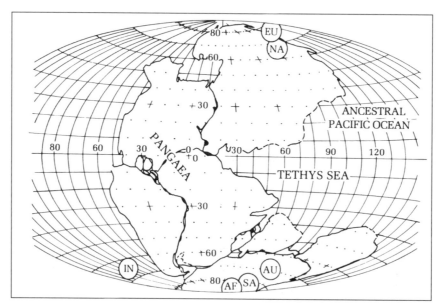

Figure III-3-2 All the continents assembled to form Pangaea, the universal landmass, as it exist-
ed at the end of the Permian Period 245 million years ago.
Source: R. S. Dietz and J. C. Holden, Journal of Geophysical Research *75, p. 4943, Copyright 1970.*
American Geophysical Union.

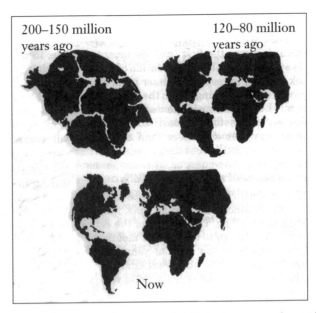

Figure III-3-3 Breakup of Pangaea and continental drift in the western hemisphere over the past 200 million years, leading to the present geography.
Source: B. Van Diver, Roadside Geology of New York, p. 27, Copyright 1997, Bradford B. Van Diver.

Nearly 100 years ago a German scientist realized that the continents had once been in different locations. That scientist was Alfred Wegener. And in 1912, when he told other scientists about this idea—which he called Continental Drift—most of them didn't accept it because no one knew how continents could move. In fact, as late as 1971 university courses were taught by scientists who knew almost nothing about Continental Drift!

About ten years after World War II ended, airborne scientists flying above the Atlantic Ocean floor and using military technology discovered regular changes in the magnetism of rocks. In a detective story that rivals Sherlock Holmes, they figured out how the continents were moving! Their airplanes were equipped with sensitive magnetometers that had been developed during the war. Magnetic changes like the ones they observed could only be caused by undersea eruptions of volcanic lava, but at that time ocean lava had only been found in Iceland.

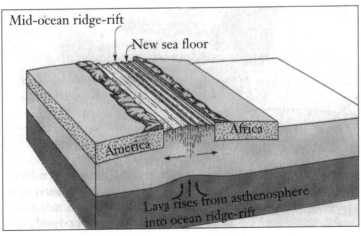

Figure III-3-4. Mid-ocean rift in the Atlantic Ocean floor showing magnetic stripes of lava with magnetic changes. This rift is also a DIVERGENT PLATE BOUNDARY.
Source: F. Press and R. Siever, Earth, 1978, Fig. 1-15, pg. 19; modified from "The Breakup of Pangaea," by R. S. Deith and J. C. Holden. Copyright 1980, Scientific American, Inc. All rights reserved.

The earth is a big magnet, and iron minerals in lava sometimes act just like iron filings sprinkled on a paper above a magnet. The minerals align themselves like little tiny magnets so that they point north and south toward the earth's magnetic poles. This especially happens in lava that has cooled quickly in sea water. Regardless of what happens, the minerals in the lava will always "remember" the north and the south end of the earth's magnet—as it was when the lava cooled.

The scientists in their airplane knew that minerals in lava line themselves up—and they also knew something else. They knew that every few thousands of years the earth's magnetic poles flip, and magnetic north and south directions change, or reverse. Minerals in new lava reflect these reversals by turning themselves in the opposite direction to match the earth's poles. It was this reversal of minerals that the scientists were seeing on their sensitive magnetometer as they flew over the ocean bottom. They realized that they might be flying over successively older eruptions of lava that showed these changes in magnetic direction. This told them that if the rock beneath their airplane was lava there was *lots* of it. The magnetic stripes of lava in the Atlantic Ocean floor are shown in Figure III-3-4.

The scientists took an oceanographic research ship out to the area they had flown over, and when they brought up volcanic lava from the ocean floor they were very excited because they suspected volcanism had something to do with why the continents were moving. They found that the lava was erupting from a long crack in the Atlantic Ocean floor called the Mid-Atlantic Ridge. (See Figure III-3-4.) The ridge became a new area of study, and soon scientists discovered that the volcanoes on the island of Iceland were right over the crack! More study showed that some force underneath the Atlantic was moving the ocean floor apart, creating a deep crack, or rift, and lava was pouring out of the crack. The lava was actually creating new sea floor! There was a lot of scurrying about in the scientific community as oceanographers rushed to measure the age of rocks on the world's ocean floors. What they found was truly exciting. The ocean floor was far younger than rock on the continents! This could only mean that new ocean floor was being created at the site of these cracks or rifts and that the whole ocean floor was *moving* away from the crack. Later, this became known as "sea-floor spreading." Satellites have since shown that both the ocean floor *and* the continents are moving. Like the ancient myth that the earth is carried on the back of a giant tortoise, the continents are being carried along by tectonic plates.

Take another look at the tectonic plates in Figure III-3-1, and you can see that the continents on either side of the Atlantic Ocean, North America, and Europe are being carried apart, but on the Pacific side of the United States, the tectonic plates and continents are colliding. When the tectonic plates bump into each other, their edges get munched. To help understand how such dramatic things can happen on the deceptively quiet surface of our planet, let's look at a hard-boiled egg. (See Activity II, Egg Tectonics.) Imagine that the shell of the egg is the earth's crust (where we live and go to school and play basketball). The crust, which includes both the continents and the ocean floor, "floats" on the mantle, which is represented by the white of the egg. The yolk of the egg is the core of the earth. The earth's crust is thin like the eggshell, and it is pretty easy to see how pieces can slide away from, against, and even underneath each other.

There is an additional concept that will help us understand things like earthquakes and volcanoes. The crust of the earth can be thought of as composed of two materials—a thicker, lighter one (the *continents*) and a thinner, heavier one (the *ocean floor*). The continents and the ocean floor "float" on the hot, slowly moving part of the upper mantle, called the *asthenosphere*, from the Greek word meaning "weak." The thinner and heavier ocean floor material tends to crack and slowly slide under things, like itself, or under the continents, sometimes dragging parts of the continental mass with it. This can cause earthquakes and volcanoes. The earthquakes happen as the ocean floor part of the plate slides downward; volcanoes are the melted crust bubbling back up to the surface. Sometimes the volcanoes will be the type that explode—and Mount St. Helens in Washington State is one of these! (You will read about Mount St. Helens in the next section.) This situation is called a *subduction zone*. (See Figure III-3-5A.)

When the tectonic plates munch together two other things can happen:

(Be sure to do Graham Cracker Tectonics, Activity III, to help visualize this)

1. The plates will slide past each other, and at the edges there will be lots of cracks and earthquakes especially where the plate contains a continental mass. This is happening in California along the San Andreas fault zone; the Pacific plate is dragging part of California toward the north. The edge of the Pacific plate is actually a zone made of lots of cracks. The zone is called a transform plate boundary (See Figure III-3-5B.)

2. The plates will butt against each other and both of their edges will rumple and crumple up into mountains and melt into each other. This type of collision is called a convergent plate boundary (See Figure III-3-5C). The former continent of India is crunching into Asia and making the Himalayan Mountains along a boundary like this one.

Today, in the United States, most of the action is on the West Coast, but let's come back to the New England coast about 500 million years before Grampa's Uncle Cyrus built their stone wall. The rocks in Grampa's stone wall record three collisions with continents that ground into the East Coast of the United States and munched New England. At first, it was a great place to take a swim on the beach; Adam's sandy beach lay along the coast. But then across the ancient ocean called Iapetus came a group of volcanic islands that didn't stop when they got to Boston (Boston wasn't there yet, of course). These islands, along with volcanic rocks from the ocean floor, very slowly plowed into New England and today we can find pieces of them. The rocks have been heated and tortured in that hot, squeezed place deep under the colliding land masses.

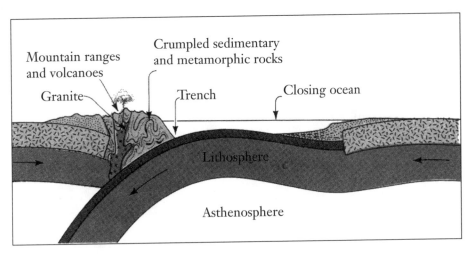

Figure III-3-5 Diagrams showing Main Types of Tectonic Plate Boundaries
A) Diagram of SUBDUCTION ZONE type of Convergent Plate Boundary. This is also how the New England coast may have looked 400 million years ago, just before the humongous collision with Avalonia.
Source: Modified from: Bunji Tagawa, Copyright March 1972 Scientific American, Inc. All rights reserved.

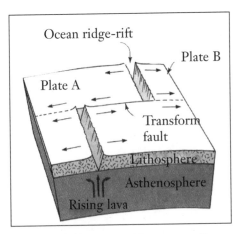

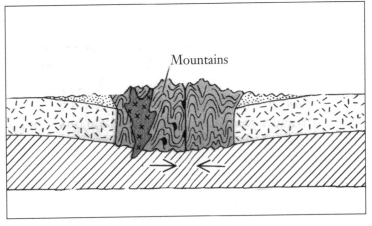

B) Diagram showing a TRANSFORM PLATE BOUNDARY. Source: F. Press and R. Siever, Earth, 2nd Ed., Fig. 1-19, p. 22, Copyright 1978, W.H. Freeman & Co.

C) Diagram showing a Convergent Plate Boundary.
Source: Modified from R. S. Dietz, "Geosynclines, Mountains, and Continent-building," Copyright March 1972, Scientific American, Inc. All rights reserved.

The second collision was a humongous one with a *big* continent called Avalonia. Avalonia was riding along on the plate that included the ocean floor of the ancient sea, Iapetus. Iapetus gradually closed, and for millions of years, the plate carrying Avalonia moved toward what is today North America. (See Figure III-3-5a.) This took a very, very long time and raised up some really big mountains.

In the third continental collision, the North American plate, along with both Africa and Europe, became part of Pangaea (see Figure III-3-2), a huge super-continent. After that, the tectonic plates moved away from each other leaving part of Europe still attached to northern New England. It was during the breakup of Pangaea that the Atlantic Ocean was born!

Activity I

Balloon Tectonics

OBJECTIVE: How far have the modern continents moved since they were parts of Pangaea?

> National Science Education Content Standard(s), K-4, 5-8
> Major Emphasis: A, E; Minor Emphasis: B, C, H;
> Other Content: Geography, Mathematics

MATERIALS NEEDED:
Several globes showing continents and oceans on the earth
Sturdy tracing paper
At least 10 pieces of sticky-sided Velcro
Balloons that will blow up to the same size as the globes

PROCEDURE:
1. Using sturdy tracing paper, trace the continents off the globe you will be using for the activity. Do Europe and Asia as one large landmass, "Eurasia," but do India separately. Locate five to eight big cities on each continent, and mark them on the traced continents.
2. Cut the continents out, label them with their name and the approximate latitude and longitude of the center. Put one piece of Velcro on the back of each and set them aside.

Note: Latitude lines go around the middle of the globe ("lat" rhymes with "fat"); the equator is at 0° latitude, the North Pole is at 90°N and the South Pole is at 90°S. Longitude lines go up and down, perpendicular to the equator. Half the globe is 0-180° E longitude, and half is 0-180°W longitude. (See Figure III-3-6.)

3. Blow up a balloon to the same size as the globe. Mark the equator, North and South Poles, and at least four major longitude lines (e.g., 0° through Greenwich, England, 90°E, 90°W, and, 180°E /180°W).
4a. Lay the cut-out continents out on a table and fit them together to make Pangaea (see Figure III-3-2).
4b. Optional: Lay the cut-out continents out on a table and fit them together to make Gondwanaland. Note: you will need to have a reference book, or search the web for a good Gondwanaland illustration.
5. Referring to Figure III-3-2, locate the approximate latitude and longitude of the center of North America as it existed in Pangaea, and place a mark on the balloon.
6. Locate the modern positions of the continents on the balloon, and Velcro them in place
7. See if you can measure how far the center of the land mass of North America has moved since it was part of Pangaea.
8. Optional—appropriate for advanced students: Using Figure III-3-1, draw all the modern plate boundaries on another balloon. Put an arrow on each plate to show how it is moving.
Note: For additional tectonic activities, visit the U.S. Geological Survey website: http://www.usgs.gov/education/

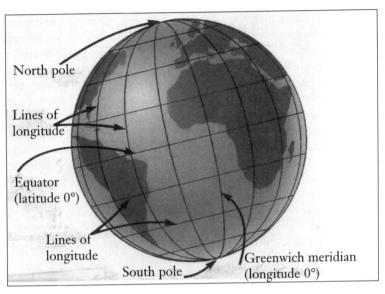

Figure III-3-6 The earth, showing lines of latitude and longitude.
Source: *F. Watt,* Planet Earth, *p. 44, Copyright 1991, Usborne Publishing Ltd.*

ACTIVITY II

Egg Tectonics (Weisgarber, 1994)

OBJECTIVES:
To introduce the relationship of the earth's crust to other layers and the interior of the planet.
To show that the earth's crust isn't flat, but is the surface of a round planet.
To aid in visualizing what happens at tectonic plate boundaries.

> National Science Education Content Standard(s), K-4, 5-8
> Major Emphasis: A, C, E; Minor Emphasis: B;

MATERIALS NEEDED: 3 (or more) hard-boiled eggs; 3 (or more) water-based markers

PROCEDURE:
1. Gently tap the eggs repeatedly on a table while rotating them to produce cracks all around the eggs. Trace along the major cracks with a water-based marker. Gently squeeze the eggs until slight movement of the shell pieces occurs. Look for places where pieces of the eggshell separate. This area represents a divergent boundary (See Figure III-3-4).
   Note: Many divergent boundaries on the earth are hidden beneath oceans and are characterized by volcanism, earthquakes, and massive heat flow due to molten rock (magma) rising up from the mantle, which is the thick layer of rock beneath the crust. The Mid-Atlantic Ridge on the bottom of the Atlantic Ocean is an example of a divergent boundary; here the North American Plate and the Eurasian Plate are separating, causing sea-floor spreading and new oceanic crust to form.
2. Next, look for places where two pieces of eggshell are colliding. This area represents a convergent boundary (See Figure III-3-5A, 5C).
   Note: Two events can occur when plates converge. If denser oceanic crust collides with lighter continental crust, the oceanic crust will buckle under the continental crust down into the mantle. This process is called subduction and is characterized by earthquakes, rock deformation and volcanism. The volcanic Cascade Range of the Pacific Northwest was formed by subduction of the Juan de Fuca Plate under the North American Plate. (See next section, Volcanoes, Where and Why?) If two equally

dense continental crusts collide, both plates will resist being subducted. In this process, the continental crust folds and deforms into a mountain range. The Himalayas are an example of a mountain-building episode which began 25 million years ago and is still occurring today as India travels northward, colliding with Asia.

3. Finally, look for places where one piece of eggshell slides past another. This area represents a transform boundary. (See Figure III-3-5B.)

   Note: The crust is not destroyed here as it is at a convergent boundary, nor is crust created as it is at a divergent boundary. As the two plates slide past each other, earthquakes occur. The San Andreas fault in California is an example of this type of boundary.

   After this experiment, use the eggs to illustrate the layers of the earth. Cut the egg in half. The shell represents the crust. The thick egg white represents the mantle. The egg yolk represents the core.

4. Optional—appropriate for advanced students: Match the measurements of thickness of the egg layers (shell, white and yolk) to the earth's layers. How thick are the real core, mantle, and crust?

ACTIVITY III

Graham Cracker Tectonics (O'Neil, 1997)

OBJECTIVE: What changes take place when the plates of the earth's crust move?

> National Science Education Content Standard(s), K-4, 5-8
> Major Emphasis: A, C, E; Minor Emphasis: B;
> Other Content: Language Arts; Snack

MATERIALS NEEDED: Two graham crackers; frosting; wax paper; water; and a cup

PROCEDURE:

1. On a piece of wax paper, spread frosting approximately 1/2 inch (1 centimeter) thick.
2. Break one of the graham crackers in half and place the two halves on the frosting leaving a small space between them (see Figure III-3-7, Diagram 1). Gently push each half of the graham cracker down and away from the crack .
3. Record your observations in the data table and answer Questions a to e.
   a) What do the graham crackers represent?
   b) What does the frosting represent?
   c) Describe what happened when you added pressure to the graham crackers.
   d) What geologic feature does this represent on the earth?
   e) What type of plate boundary does this represent? Explain your choice.
4. Take the second graham cracker and break it into four sections
5. Place the ends of two sections in the cup of water for about five seconds, until they are soft, but take them out before they fall apart.
6. Put them on frosting layer on the wax paper so the wet ends are together (see Figure III-3-7, Diagram 2). Now gently push them toward each other. Record your observations in the data table and answer Questions f to h.
   f) Describe what happened when you pushed the graham crackers together.
   g) What geologic features do the ends of the graham crackers represent?
   h) What type of plate boundary does this represent? Explain your choice.
7. Place the remaining two sections of graham crackers next to each other (see Figure III-3-7, Diagram 3). Gently push them together, and slowly slide the sections past each other.
8. Record your observations in the data table and answer Questions i and j.
   i) Describe what happened when you moved the graham crackers.
   j) What type of plate boundary does this represent? Explain your choice.
9. Time to eat your plates!

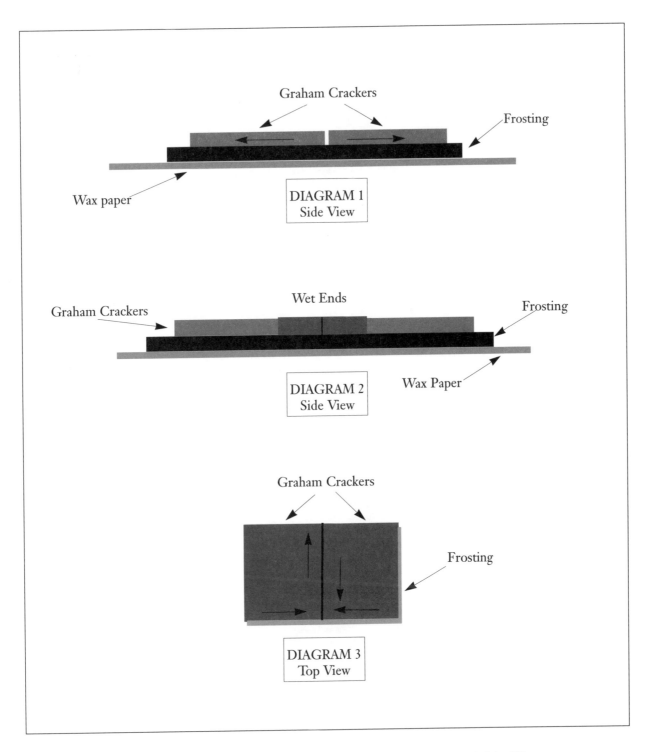

Figure III-3-7 Diagrams Illustrating Graham Cracker Tectonics, Activity III

Graham Cracker Tectonics Data Table		
Procedure Step Number	Type of Plate Boundary	Describe the changes that took place
2 Diagram 1		
6 Diagram 2		
8 Diagram 3		

<u>ACTIVITY IV</u>

How to Determine True North
(Adapted from Hansgen, 1996)

OBJECTIVE: Can you mark on the ground a true north-south line?

> National Science Education Content Standard(s), K-4, 5-8
> Major Emphasis: A, B, C, E; Minor Emphasis: H;
> Other Content: Measurement, Geography

MATERIALS NEEDED FOR EACH GROUP:
    A horizontal surface (such as a sidewalk)
    Support stand and rod (you can use curtain rods, dowel sticks, or a standard laboratory support stand with rod)
    2 large sheets of paper
    Masking tape
    Meter or yard stick
    Watch
    Sharp pencil
    Teacher also needs a scout or army-type compass and a large protractor.

PROCEDURE:
1. A straightforward method to lay out a north-south line is to first determine midday. Midday is simply that time halfway between the times of sunrise and sunset, which can be ascertained by calling the local television station or newspaper, watching the local evening news, or, for the official source of time used in the United States, check the U.S. Naval Observatory website: http://tycho.usno.navy.mil/ The times given will be accurate to within a couple of minutes, depending on your location with respect to the station.
2. The class can be divided into small groups. Each group will need the items listed above under materials.
3. Take the class outside on a sunny day about a half hour before your calculated value for midday.
4. Find a horizontal surface, such as a sidewalk, on which to work.
5. Place the support stand (see materials) on the level surface and arrange the large sheet of paper so that the shadow of the tip of the rod falls on the paper.

6. Tape the paper to the level surface. (Remember, the shadow is going to move; that is why we have an extra piece of paper.)

7. The position of the base of the stand should be outlined by pencil on the paper so that if the stand gets bumped, it can be returned to its original position.
Note: To add a worthwhile flair to this experiment, the instructor can set up a similar apparatus in the morning and mark positions of the tip of the shadow every half hour. Record the times on the paper near the appropriate mark. The teacher can then connect these points with a smooth curve, providing a dramatic representation of how far the shadow (or "sun") moves in just a couple of hours.

8. Students should begin their measurements at least 20 minutes before the calculated value for midday and should measure the length of the shadow every 2 minutes for approximately 40 minutes.

9. They should mark the position of the tip of the rod's shadow and note the time beside each mark. Measurements should continue until the length of the shadow begins to noticeably increase.
Note: Midday is when the length of the rod's shadow is the shortest. A line drawn on the paper between the mark representing the tip of the minimum shadow and the center of the base of the support rod provides a true north-south line.

What the students are witnessing is empirical evidence that the earth is rotating about its axis. They are watching the earth spin. One should keep in mind, though, that this is evidence, not proof. A stationary earth about which the Sun makes a daily orbit provides an alternative explanation. A field trip to a local science museum where a Foucault pendulum is on display would provide more direct evidence that the earth is rotating.

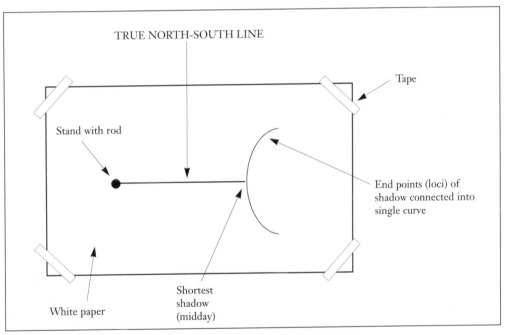

Figure III-3-8. Diagram of set-up for determining a line pointing to true north.

10. With the aid of the compass and the protractor, students can now determine the angle of declination (or variation) between magnetic north, the direction the compass needle points, and the line which represents true north they have drawn on the paper.

11. They should compare this angle of declination with the accepted value for the angle of magnetic declination (or variation) given on a U.S. Geological Survey topographic map of their area (don't forget to correct the magnetic declination from the value given on the map to the date of your activity). If this experiment is done with moderate care, the students will be pleased with the comparison.

FURTHER QUESTIONS AND EXPERIMENTS COME IMMEDIATELY TO MIND:
- How did the measured time for midday compare with the computed time?
- Why wasn't midday at noon?
- What were sources of error in this experiment and how might they be minimized?
- For any given time, will the length of the shadow change from day to day? If so, why? On what date will shadows be shortest? Longest?

Shadow experiments demonstrate that excellent science can be done with simple apparatus. After all, the old Greeks did remarkably well with what nature provided: a stick, the sun, and some human ingenuity!

REFERENCES AND FURTHER READING:

Gilman, Richard, and C.A. Chapman, T.V. Lowell, and H.W. Borns, Jr.,1988. THE GEOLOGY OF MOUNT DESERT ISLAND, A VISITOR'S GUIDE TO THE GEOLOGY OF ACADIA NATIONAL PARK: U.S. Dept. of the Interior/Maine Geological Survey, Dept. of Conservation, U.S., 50 pp.

Glenn, William, 1975. CONTINENTAL DRIFT AND PLATE TECTONICS: Charles E. Merrill Publishing Company, Columbus, OH, 188 pp.

Hansgen, Richard 1996. HOW TO DETERMINE TRUE NORTH: Hands On Earth Science, No. 9, Ohio Department of Natural Resources, Division of Geological Survey, Columbus, OH. Originally published in: The Physics Teacher, 1995, v. 33, p. 116-117.

McPhee, John, 1993. ASSEMBLING CALIFORNIA: Farrar, Straus and Giroux, New York, NY, 304 p.

O'Niel, T., 1997. GRAHAM CRACKER TECTONICS: E.S.P.I.R.I.T., New York.

Raymo, Chet and M.E. Raymo, 1989. WRITTEN IN STONE—A GEOLOGICAL HISTORY OF THE NORTHEASTERN UNITED STATES: The Globe Pequot Press, Old Saybrook, CT, 168 pp.

Van Diver, Bradford, 1997. ROADSIDE GEOLOGY OF NEW YORK: Mountain Press Publishing Company, Missoula, MT, 411 pp.

Van Diver, Bradford, 1995. ROADSIDE GEOLOGY OF VERMONT AND NEW HAMPSHIRE: Mountain Press Publishing Company, Missoula, MT, 230 pp.

Weisgarber, Sherry L., 1994, EGG TECTONICS: HANDS-ON EARTH SCIENCE, No. 2, Ohio Department of Natural Resources, Division of Geological Survey, Columbus, OH.

Wilson, J. Tuzo (Introductions), 1970. CONTINENTS ADRIFT: W.H. Freeman and Company, New York, NY, 172 pp.

USEFUL WEBSITES:

Magnetic Stripes and Isotopic Clocks
•http://pubs.usgs.gov/publications/text/stripes.html
U.S. Geological Survey site about the discovery of sea-floor spreading.

This Dynamic Earth
•http://pubs.usgs.gov/publications/text/dynamic.html
The Story of Plate Tectonics—complete book published by the U.S. Geological Survey.

Active Tectonics
•http://www.muohio.edu/tectonics/ActiveTectonics.html

The ABC's of Plate Tectonics
•http://home.earthlink.net/~dlblanc/tectonic/ptABCs.html
by Donald L. Blanchard.

Earthquakes & Plate Tectonics
•http://wwwneic.cr.usgs.gov/neis/plate_tectonics/rift_man.html
USGS National Earthquake Information Center explains the connection.

Earthquake Information
•http://wwwneic.cr.usgs.gov/neis/bulletin/97_EVENTS/97_EVENTS.html
National Earthquake Information Center, U.S. Geological Survey, provides a wealth of information including quakes happening now.

Introduction to Plate Tectonics
•http://www.hartrao.ac.za/geodesy/Plate.html
Covering the chemical and physical layers of the Earth, historical development of the theory, and descriptions of the location and types of plate boundaries.

Plate Tectonics [geocities.com] —Students' Site
•http://www.geocities.com/CollegePark/Quad/7635/tec-over.html
Providing basic views/some in-depth terms and facts.

Tectonic Plate Motion
•http://cddis.gsfc.nasa.gov/926/slrtecto.html

Mountains
•http://aleph0.clarku.edu/rajs/nepal-him.html
Comparing Himalayan to Rocky Mountains.

Tectonics of the Eurasian Plate
•http://wcuvax1.wcu.edu/~cm900/sam.html
Generalized overview on the convergent, divergent and transform margins.

III-4
Volcanoes, Where and Why?

OBJECTIVES:
1. To show that volcanoes often occur where continents crunch together.
2. To show that like a shaken up bottle of Coca-Cola, volcanoes are powered by gas pressure.
3. To show that the process of volcanism gave us the crust of the earth and our atmosphere.

> *"Earthquakes rumbled, thrusting the mud into mountains. Volcanoes fumed. The smell of sulfur—like rotten eggs—spewed into the air."*
>
> From *Stone Wall Secrets*

One of our country's geologists was killed when Mt. St. Helens exploded. David A. Johnston, volcanologist with the U.S. Geological Survey, was six miles away at 7:00 AM on May 18, 1980, and had just radioed in measurements of how much the volcano was bulging. Nothing unusual. Despite careful monitoring for many months, the timing and size of the eruption caught David and everyone else off guard.

For more than 40,000 years Mt. St. Helens has belched forth many, many tons of ash. Just a few days after the May 1980 explosion, ash from Mt. St. Helens covered one-quarter of the United States! (See Figure III-4-1.) Where did it all come from? Would you believe that the ash falling on Nebraska came from the floor of the Pacific Ocean? It took a very long time to get there, and it traveled underground!

How can this happen? Take a look at the map of tectonic plates in Figure III-4-2. Can you find the United States? Now look for the Juan de Fuca plate off the West Coast. The Juan de Fuca plate is sliding down under northern California, Oregon, and Washington. As it descends, the forward edge melts and the magma rises up to create Mt. St. Helens, Mt. Ranier, and the string of volcanoes in the Cascade Mountains. To help you understand this volcanic "plumbing," look at the diagram in Figure III-4-3. Can you find two areas where magma comes up to the surface? Look at the Juan de Fuca Ridge out in the Pacific Ocean—and at Mt. Ranier. Magma comes up in both places.

If you are beginning to think that volcanoes are part of a huge earth-surface recycling program, you are right! As ocean plates made of volcanic rock slide beneath continents and melt again, new magma rises. And this is why volcanoes are important to life on earth. Magma and ash bring us the ingredients for new soil to grow our food, and gases for our atmosphere. These things are happening very slowly, of course, but they are happening somewhere on the earth nearly every day, and have been for billions of years.

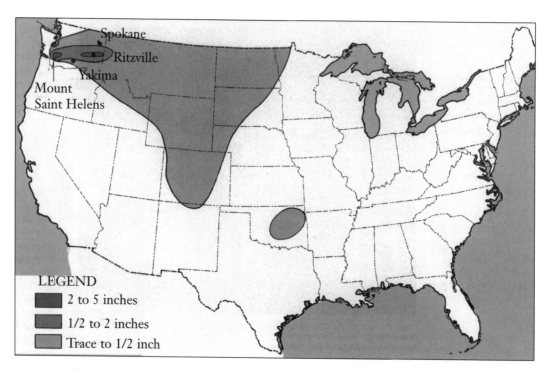

Figure III-4-1. Distribution of ash fallout from the May 18 eruption of Mount St. Helens. Up to a half inch of ash fell over the largest area. From which direction was the prevailing wind?
Source: R. Tilling and others, Eruptions of Mount St. Helens: Past, Present, and Future, p. 18, 1990, Department of the Interior/U.S. Geological Survey.

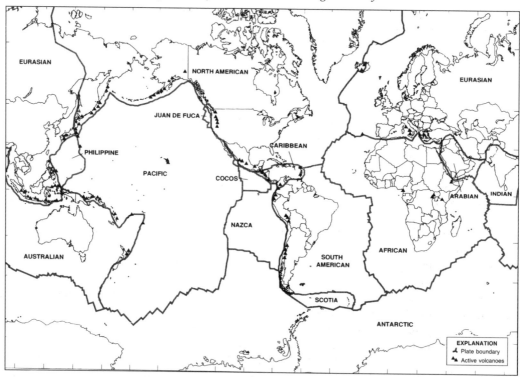

Figure III-4-2. Tectonic plates of the world showing what is left of the Juan de Fuca Plate along the West Coast of the United States. The Juan de Fuca Plate is a piece of the ocean floor that is sliding down under the states of Washington and Oregon. Magma from the melted plate feeds volcanoes like Mt. St. Helens.
Source: S. Brantley, Volcanoes of the United States, p. 7, 1994 Department of the Interior/U.S. Geological Survey.

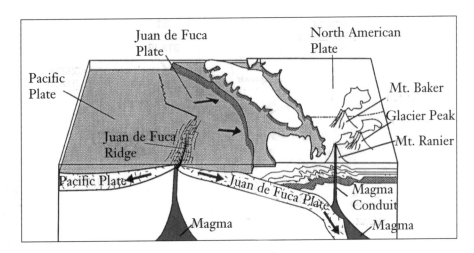

Figure III-4-3. Source of magma for Mt. St. Helens. In the Pacific Northeast, the Juan de Fuca Plate plunges beneath the North American Plate. As the denser plate of oceanic crust is forced deep into the earth's interior beneath the continental plate, a process known as subduction, *it encounters high temperatures and pressures that partially melt solid rock. Some of this newly formed magma rises toward the earth's surface to erupt, forming a line of volcanoes along the subduction zone.*
Source: S. Brantley, Volcanoes of the United States, *p. 8, 1994, Department of the Interior/U.S. Geological Survey.*

So why did Mt. St. Helens explode? Because it was filled with steam just like a pressure cooker! Red hot magma rising under the mountain forced it apart and groundwater (rain that had sunk into the ground) rushed through cracks into the hot magma. The water was instantly changed to steam, and like a shaken bottle of Coca-Cola, it created enormous pressure against the walls of cracks and fissures inside the volcano. Earthquakes rumbled as magma continued to crack the rocks, and steam pressure increased. The volcano bulged as it filled up with magma and steam. It was this bulge that David Johnston was measuring when, on May 18, one of the earthquakes caused an enormous landslide that took away the cover over the pressure-cooker below. It was like popping the cork out of a bottle of champagne. The whole side of the mountain blew out with awesome force.

What would you do if a volcano erupted near your town? Well, if the volcano was like Mt. St. Helens, there wouldn't be much you could do to save yourself and your family. The blast came zooming across the countryside at supersonic speed. There were tons of ash, and it was hot. Hot air mixed with ash can travel down the sides of volcanoes at speeds of hundreds of miles an hour. Towns can be wiped out in the wink of an eye. And then there were the huge landslides and sediment-flooded valleys. Hot ash mixed with melting snow quickly buried homes, roads, and people under tens of feet of mud. In the Hawaiian Islands, volcanoes are much quieter about their eruptions. There are no huge explosions, just lava flowing relentlessly—it can still destroy homes and villages, just not so fast. The difference in volcano behavior has to do with the kind of magma. And that is a story for another time.

In *Stone Wall Secrets*, Grampa imagines the sulfur smell ("like rotten eggs") of volcanic gases spewing into the air. Why do volcanoes smell like rotten eggs? It is because gases belching out of the earth include sulfur dioxide, or SO_2, and hydrogen sulfide, or H_2S. Yes, there is sulfur in eggs, and when eggs spoil, the sulfur makes hydrogen sulfide gas.

What does the size of the crystals in a rock tell us about how fast the rock cooled? Some volcanic lava has a composition very similar to granite (quartz, feldspar and "dark minerals"). But rocks formed from cooling lava don't look like granite. Remember that granite has a coarse texture—the crystals of different minerals are easily big enough to see. And where does granite form? Granite forms from magma that creeps into cracks deep in the earth where it cools so slowly that mineral crystals have a chance to grow larger. Lava, on the other hand, cools so fast that the rock gets hard before crystals can get large. Lava-derived rocks such as basalt and rhyolite almost always have crystals, but they are very tiny. Remember that at the earth's spreading center a great deal of lava flows out into the cold ocean water. And there are no crystals at all in obsidian and pumice, especially when they are chilled in seawater.

ACTIVITY I

Draw a Subduction Zone with Volcanoes

OBJECTIVES:
1. To show that volcanoes occur where continents crunch together.
2. To help your students learn an extremely important geologic process.

> National Science Education Content Standard(s), K-4, 5-8
> Major Emphasis: A, C, E; Minor Emphasis: G
> Other Content: Art

MATERIALS NEEDED: Colored pencils; felt tip markers, etc.; a ruler; and paper.

PROCEDURE: Have your students draw, color and label the diagram in Figure III-4-3.

ACTIVITY II

Build a Volcano

OBJECTIVE: To show that volcanoes are powered by gas pressure.

> National Science Education Content Standard(s), K-4, 5-8
> Major Emphasis: A, C, E; Minor Emphasis: B, G;
> Other Content: Art

MATERIALS NEEDED: Pre-mixed wallpaper paste; chicken wire; newspaper or computer printout paper; 12-oz plastic glass; vinegar; baking soda; alum; measuring spoons; red food coloring.

PROCEDURE:
1. Construct a volcano out of papier-mâché. Use a picture of an active volcano, and try to stay as accurate as you can. Build the volcano around a plastic tube large enough to hold a plastic glass with the volcanic "lava" mixture in it. Use chicken wire to make the shape of the mountain. Wet the paper with the wallpaper paste and build up several layers on the chicken wire. Be creative: rocks and other features can be made out of paper thoroughly wet with paste, then crunched up and sculpted to the shape you want. Let the whole thing dry completely, then paint it to look realistic.
2. Mix the ingredients for the "eruption": 2 tbs. vinegar, $1/2$ tbsp. baking soda, $1/2$ tsp. alum, red food coloring.
   Put this mixture into a 12-ounce plastic glass, or other container that will fit just inside the top of the volcano. When you're ready for your volcano to erupt, stand back and add $1/2$ teaspoon more alum. *Note:* You might want to do this activity outside, or in a large sink. This mixture is safe to touch, but is not edible! Be sure you protect your table or floor under the volcano with plastic, to make clean-up easier.

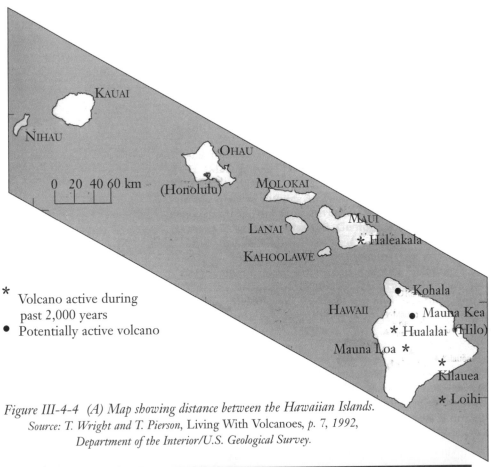

* Volcano active during
  past 2,000 years
• Potentially active volcano

Figure III-4-4 (A) Map showing distance between the Hawaiian Islands.
Source: T. Wright and T. Pierson, Living With Volcanoes, *p. 7, 1992,*
Department of the Interior/U.S. Geological Survey.

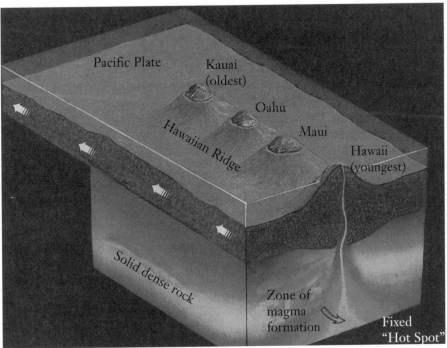

(B) Block Diagram showing the Hot Spot under the Hawaiian Islands.
Source: R. Tilling and others, Eruptions of Hawaiian Volcanoes: Past, Present, and Future, *p. 12,*
1993, Department of the Interior/U.S. Geological Survey.

ACTIVITY III

Discovering Movement of the Earth's Tectonic Plates

OBJECTIVE: To use the chain of volcanic Hawaiian Islands to show movement of the Pacific Plate.

> National Science Education Content Standard(s), K-4, 5-8
> Major Emphasis: A, C, E; Minor Emphasis: B, G;
> Other Content: Mathematics, Measurement, Geography

INFORMATION NEEDED:

Major Hawaiian Islands	Average Ages, in MILLIONS of years
— Kauai	4.7
— Oahu	2.8
— Molokai	1.5
— Maui	1.1
— Hawaii	0.7

Islands	Distance BetweenIslands, in kilometers	Speed of the Pacific Plate, in kilometers per MILLION years	Speed of the Pacific Plate, in centimeters per year
Kauai to Oahu	160	84	8.4
Oahu to Molokai	100	77	7.7
Molokai to Maui	80	200	20.0
Maui to Hawaii	140	350	35.0

MATERIALS NEEDED: Calculator; diagrams in Figure III-4-4.

PROCEDURE:

1. Determine the distance between the centers of each of the five major Hawaiian Islands. Use the map in Figure III-4-4A. Cut a piece of paper about three inches long and one inch wide. Lay the edge of the paper along the scale [0 - 60 kilometers (km)] and mark off the "0," "20," "40," and "60" kilometer distances. Use your "ruler" to measure the distance between the islands.

2. Calculate the speed of the Pacific Plate between each island.
   Example:
   • Distance between Kauai and Oahu = 160 km
   • Difference in the ages of Kauai and Oahu = 1.9 million years (4.7 - 2.8 million years)
   • Speed of Pacific Plate = 160 km divided by 1.9 million years = 84 km/million years
   Answer: 8.4 centimeters per year.
   Note: The answers are given in the table above, and you can have your class draw the islands and put the speeds on their map.

3. Using the block diagram in Figure III-4-4B have a class discussion on hot spots under the Hawaiian Islands and what they mean.

REFERENCES AND FURTHER READING:

Brantley, Steven, 1994. VOLCANOES OF THE UNITED STATES: U.S. Dept. of the Interior/U.S. Geological Survey, U.S. Government Printing Office, Washington, DC, 44 pp.

Simkin, T., and L. Siebert, 1994, VOLCANOES OF THE WORLD 2ND EDITION: Geoscience Press in Association with the Smithsonian Institution's Volcanism Program, Tucson, AZ, 368 pp, 33 illustrations, 50 color maps.

Tilling, Robert, C. Heliker, and T. Wright, 1993. ERUPTIONS OF HAWAIIAN VOLCANOES: PAST, PRESENT AND FUTURE: U.S. Department of the Interior/U.S. Geological Survey, U.S. Government Printing Office, Washington, DC, 54 pp.

Tilling, Robert, L. Topinka, and D. Swanson, 1990. ERUPTIONS OF MOUNT ST. HELENS: PAST, PRESENT AND FUTURE: U.S. Department of the Interior/U.S. Geological Survey, U.S. Government Printing Office, Washington, DC, 56 pp.

Tilling, Robert, 1992. VOLCANOES: U.S. Department of the Interior/U.S. Geological Survey, U.S. Government Printing Office, Washington, DC, 45 pp.

Wright, T. L. and T. C. Pierson, 1992. LIVING WITH VOLCANOES: U.S. Geological Survey Circular 1073, Federal Center, Box 25425, Denver, CO 80225, 57 pp.

USEFUL WEBSITES:
From the U.S. Geological Survey:

Frequently Asked Questions About Volcanoes
•http://geology.er.usgs.gov/eastern/volcanoes/volquest.html

A Paper Volcano Model that Students Can Make
•http://www.usgs.gov/education

About Hawaiian Volcanoes
•http://hvo.wr.usgs.gov/
This USGS site from the Hawaiian Volcano Observatory offers something for everyone. From fun activities for kids to learn about volcanoes, to a volcano watch for the latest Hawaiian volcanic eruptions, to how volcanoes work.

The Electronic Volcano
•http://www.dartmouth.edu/~volcano/index.html
•http://www.dartmouth.edu/pages/rox/volcanoes/elecvolc.html
This site is a "front door" for links to data sets and images of volcanoes around the world (Press and Siever, 1998). From Dartmouth College, Dept. of Earth Sciences. Offers visual information, maps of active volcanoes, current events, and research. Offers material in several languages. A primary source of information about active volcanoes.

Hawaii's Center for Volcanology
• http://www.soest.hawaii.edu/GG/hcv.html
Maintained by the University of Hawaii (Manoa), this site is a good source of information on Hawaiian volcanoes.

Teaching About Hawaiian Volcanoes
•http://volcano.und.nodak.edu/vwdocs/vwlessons/atg.html
A teacher's guide to the geology of Hawaii Volcanoes National Park. Also great source of information about plate tectonics.

Teaching about Active Alaskan Volcanoes
•http://www.avo.alaska.edu/
Alaska Volcano Observatory site is rich source of teaching material.

Volcano World
•http://volcano.und.nodak.edu/vw.html
This site, at the University of North Dakota, is a "front door" for almost everything available on volcanoes. Features include current volcanic events and "Ask a Volcanologist."

Catalog of Active Volcanoes
•http://www.dartmouth.edu/~volcano/EvA.html

Teaching About Volcanoes
•http://www.athena.ivv.nasa.gov
An Athena site.

Does the Earth Itch?

OBJECTIVE: To show how weathering (the interaction between the surface of the crust and the atmosphere) brought up the stones for Grampa's wall.

A biologist would tell us that before the earth could itch it would have to have skin and a brain. Well, I think the earth *does* have skin—it's not like our skin, of course—it is the outer surface of the crust. As to whether the earth has a brain—I guess a philosopher would have to speak to that question. So, does the earth itch? Like the poison ivy toxins that make itchy bumps on our skin, there are many forces that carve, mold, build up, and destroy the earth's surface where we live. It gets cold, then hot; it rains and freezes; the wind blows; and rivers flood. Not to mention meteor impacts, landslides, earthquakes, and layers of volcanic ash. How does the earth *react* to these forces?

We will not be able to discuss here all of the many things that change the face of the earth, but we do want to look at one particular process because it brought Grampa the stones for his wall. The process is called *frost heave*, and Grampa tells Adam about it:

> *With the trees gone, the topsoil thinned and the ground froze deeper than ever before. Left by the glacier, stones that hadn't seen the light of day for thousands of years heaved to the surface and had to be hauled away. Like the farmers before him, Grampa had to put the stones somewhere, so he, too, stacked them onto walls.*

> From *Stone Wall Secrets*

When trees growing in the rich soils that covered New England were cleared to make the early farms, it was like taking off your overcoat on a cold day. The deep, soft, warm cover of leaves was soon gone, exposing the glacial debris underneath. The cover, or "topsoil" was plowed under, creating a rich mixture that provided years of bountiful crops: corn, beans, pumpkins, and grain. But after a while, without its soft, springy texture, the soil became a mixture of sand and clay with large and small rounded boulders and cobbles hiding in it. Rain washing across the countryside carried away the smaller grains and gradually the mixture became more concentrated in boulders and cobbles. Enter winter cold and spring melt.

In the spring, Grampa's Uncle Cyrus and his straining team of horses struggled through sticky mud to clear the fields of boulders that appeared as if by magic. A clutter of glacial boulders is shown in Figure III-5-1. In the winter the boulders and cobbles froze solid. Year after year, with the cycle of freezing and thawing, each spring brought more boulders to the surface. Eventually scientists and farmers alike have learned why the fields continue to produce a crop of rocks.

It turns out that expanding ice in the winter starts a process that raises the boulders. How can ice move something as heavy as a boulder? To help answer this question, remember that water expands as it freezes—and take a look at what happens to the boulder in Figure III-5-2. At first (A) during Spring/Summer/Fall the boulder is a few inches below the surface of the soil (you might strike it with your shovel if you were digging a hole). Freezing cold New England winter #1 (B) forms ice lenses that surround the boulder and its soil. The lenses grows larger and larger because water is drawn up to them from deep in the soil. As the ice expands, the boulder and soil around it are lifted up. The second spring thaw (C) melts the ice, and all the extra water flows downward, away from the boulder. As the water drains away, the soil settles, but the boulder will often fail to go all the way back to its original position—it might shift sideways, tilt, or some pebbles might get underneath and hold it up. Over many years (D, E, and F), the boulder will move slowly upward until one spring (G) Uncle Cyrus stubs his toe, and the boulder ends up in his stone wall.

Figure III-5-1 Pasture near Lowville, New York, cluttered with glacial erratics.
Source: B. Van Diver, Drawing by T. Jancek, Roadside Geology of New York, pg. 39,
Copyright 1985, B. Van Diver.

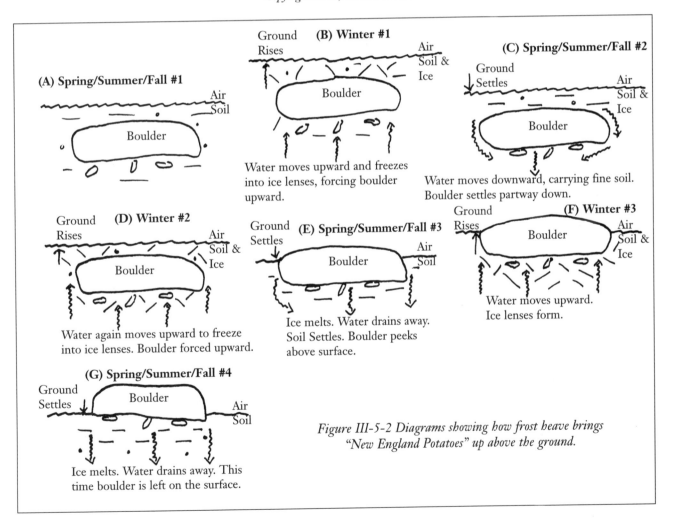

(A) Spring/Summer/Fall #1

Boulder

Air
Soil

(B) Winter #1

Ground
Rises

Air
Soil &
Ice

Boulder

Water moves upward and freezes
into ice lenses, forcing boulder
upward.

(C) Spring/Summer/Fall #2

Ground
Settles

Air
Soil &
Ice

Boulder

Water moves downward, carrying fine soil.
Boulder settles partway down.

(D) Winter #2

Ground
Rises

Air
Soil &
Ice

Boulder

Water again moves upward to freeze
into ice lenses. Boulder forced upward.

(E) Spring/Summer/Fall #3

Ground
Settles

Air
Soil

Boulder

Ice melts. Water drains away.
Soil Settles. Boulder peeks
above surface.

(F) Winter #3

Ground
Rises

Boulder

Air
Soil &
Ice

Water moves upward.
Ice lenses form.

(G) Spring/Summer/Fall #4

Ground
Settles

Boulder

Air
Soil

Ice melts. Water drains away. This
time boulder is left on the surface.

Figure III-5-2 Diagrams showing how frost heave brings
"New England Potatoes" up above the ground.

ACTIVITY I

Freezing Ice in a Plastic Bottle

OBJECTIVE: To show the power of expanding ice.

> National Science Education Content Standard(s), K-4, 5-8
> Major Emphasis: A, B, C; Minor Emphasis: E;
> Other Content: Public Speaking

MATERIALS NEEDED: Small plastic soft drink bottles for each student in your class. The bottles must have screw-on caps and a narrow neck. They should be light colored or clear.

PROCEDURE:
1. Have each student put her/his name on a bottle. Fill the bottles with clean water to within $^1/_2$ inch of the top, mark the water level, and screw on the cap.
2. Place all the bottles in a freezer overnight.
3. In the morning, have each student draw his or her bottle and write a short description of what happened. (Ice forming in the bottle will expand, might split the bottle, and/or pop the cap off and goosh out the top!)
4. Have a class discussion about why. Prompt for the concept that when water freezes, its volume increases. At room temperature, 70° F., one cubic centimeter (for conversion to English units see chart, Appendix 2) of liquid water (it would be helpful to draw one on the board) occupies one cubic centimeter. But at 0°F, the same water once frozen occupies more than one cubic centimeter.
5. How does this relate to stones popping up in Grampa's field? In the winter, ice pushes with great force on the boulder. As long as the force is equal in all directions, the boulder won't move, but as soon as the ice melts over the top of the boulder, it will move upward just like the ice in your students' bottles.

ACTIVITY II

Do Rocks Last Forever?

OBJECTIVE: To demonstrate the effects of mechanical weathering.

> National Science Education Content Standard(s), K-4, 5-8
> Major Emphasis: A, C, E; Minor Emphasis: B, G;
> Other Content: Public Speaking

MATERIALS NEEDED:
   Plaster of Paris; a small balloon; water; two empty pint milk cartons (bottom halves only); a freezer

PROCEDURE:
1. Fill the balloon with water until it is the size of a ping-pong ball. Tie a knot at the end.
2. Mix water with plaster of Paris until the mixture is as thick as yogurt. Pour half of the plaster in one milk carton and the other half in the other.
3. Push the balloon down into the plaster in one carton until it is about 1/4 inch under the surface. Hold the balloon there until the plaster sets enough so that the balloon doesn't rise to the surface.
4. Let the plaster harden for about one hour.
5. Put both milk cartons in the freezer overnight.
6. Remove the containers the next day to see what happened.

QUESTIONS:
1. What happened to the plaster that contained the balloon?

2. What happened to the plaster that had no balloon?
3. Why is there a difference?
4. Which carton acted as the control? Why?
5. How does this experiment show what happens when water seeps into a crack in a rock and freezes?
6. Does this activity suggest why streets get pot holes in the winter?

RESULTS:
The plaster containing the balloon should have cracked as the water in the balloon froze and expanded.
Explain that when water seeps into cracks in rocks and freezes, it can eventually break rocks apart!

ACTIVITY III

Shake It Up

OBJECTIVE: To demonstrate the effects of mechanical weathering.

National Science Education Content Standard(s), K-4, 5-8
Major Emphasis: A, C, E; Minor Emphasis: B, G;

MATERIALS NEEDED:
15 rough, jagged stones that are all about the same size, but are of different composition
3 unbreakable containers with lids (like coffee cans)
3 clear jars; a pen; paper; masking tape

PROCEDURE:
1. Separate the stones into three piles of five. Try to keep the stones in each pile as similar as possible. Put each pile on a sheet of paper. Optional: Weigh each pile of five dry stones. Record the weights.
2. Label each pile A, B, or C. Label each can and jar A, B, or C.
3. Fill Can A halfway with water and put in the stones from Pile A. Do the same with Can B and Pile B and Can C and Pile C. Let the stones stand in the water over night.
4. The next day, pass Can A around your class and have each student hold it in both hands and shake it hard so that it is shaken a total of 100 times.
5. Remove the stones from Can A with your hands and pour the water into Jar A. Be sure to get all the sediment in the can out into the jar. Observe the stones and the water.
6. Pass Can B around your class so that it is shaken a total of 1,000 times. Remove these stones and pour the water into Jar B. Be sure to get all the sediment in the can out into the jar. Observe·the stones and the water.
7. Do not shake Can C. Remove these stones and pour the water into Jar C. Observe the stones and the water.
8. Compare the three piles of stones and the three jars of water. Optional: Dry the stones from each pile and weigh them. Compare the weights with the starting values.

QUESTIONS:
1. How do the piles of stones differ? Why?
2. Which pile acted as the control? Why?
3. How do the jars of water differ?
4. How does this experiment show what happens to stones that are knocked about in a fast-moving river?

RESULTS:
The stones that were shaken should have more rounded edges than the stones that weren't shaken—and the stones in Can B should have rounder edges than the ones in Can A. Both jars should have some sediment in the bottom, but Jar B should have more sediment because more shakes would have broken off more bits of rock. The same thing happens to rocks that are carried along in rivers or are tumbled about by waves.

ACTIVITY IV

Steel Wool and Water

OBJECTIVE: To demonstrate the effects of chemical weathering.

> National Science Education Content Standard(s), K-4, 5-8
> Major Emphasis: A, C, E; Minor Emphasis: B, G;

MATERIALS AND EQUIPMENT NEEDED:
3 shallow dishes; water; 3 pieces of steel wool; gloves; salt

PROCEDURE:
1. Place each piece of steel wool in a shallow dish (wear gloves, because steel wool can give splinters).
2. Pour equal amounts of water over two of the pieces of steel wool. Leave the third piece dry.
3. Sprinkle one of these wet pieces with plenty of salt.
4. Observe and compare the pieces every day for a week.

QUESTIONS:
1. What happened to each piece of steel wool?
2. Which piece changed the most?
3. Why do you think the steel wool changed?
4. Which piece of steel wool acted as the control?
5. What does this experiment have to do with weathering?

RESULTS:
When iron gets wet, the water acts as an agent to speed up oxidation (oxidation occurs when oxygen combines with another substance). In this case, oxygen in the water combined with the iron in the steel wool to form an iron oxide, or rust. Rust is a weaker material than the original metal and erodes quickly. When salt is added to the water, it speeds up the oxidation of iron. So, the steel wool in the salt water should have changed the most. The same thing happens to rocks that contain iron as happens to cars during northern winters when salt is put on the roads.

IDEAS FOR OTHER ACTIVITIES:
- Measure the temperature in the same place for several months, morning and afternoon, and plot the readings on a graph.
- Design a fence post that wouldn't rise out of the ground in the spring.
- Draw a diagram of rocks coming to the surface during the freeze/thaw process.
- Try Activity IV with the mineral pyrite.

QUESTIONS FOR FURTHER STUDENT RESEARCH:
How many processes can you name that would affect the "skin" of the earth?
Why do iron meteorites develop a rusty surface?
Where does rust come from?
Why does top soil erode when the trees are gone? Where does it go? How do we know?

REFERENCES AND FURTHER READING:
Stockley, C., F. Watt, F., et al, USBORNE SCIENCE EXPERIMENTS—THE USBORNE BOOK OF THE EARTH (a bind-up of the following titles: WEATHER AND CLIMATE, PLANET EARTH, ENERGY AND POWER, AND ECOLOGY): Usborne, 192 pp., ISBN 0-7460-1454-6, $18.95.

USEFUL WEBSITES:

Resources for Teaching About Geologic Processes
•http://www.usgs.gov/education/
K-12 education material from the U.S. Geological Survey.

K-12 Activities About Sedimentary Geologic Processes
•http://www.beloit.edu/~SEPM/
A website from the Society for Sedimentary Geology.

III-6
Will The Glaciers Come Back?

OBJECTIVE: To show that glaciers covered New England; to show how this was discovered; and to show the relationship between the emergence of mankind and glaciers.

Reaching below the silken cobwebs, Grampa pulled out a smooth, black stone covered with tiny scratches. "This stone tells one of my favorite stories—the story of the great Ice Age. Twenty thousand years ago, a giant glacier oozed southward over all New England, smothering even the tallest mountains with ice. On top, the glacier was a snowflake desert: frigid, white, and blinding beneath a brilliant sun. On the bottom, the ice lay in total darkness; it pressed the brittle, broken rock back into hard mud.... The glacier streamed by for thousands of years. The climate warmed; the ice changed to muddy water and then disappeared, leaving behind a windy, treeless world.."

From *Stone Wall Secrets*

Just a few thousand years ago an ice sheet nearly *one mile thick* covered New England! As we drive on Interstate 95 now from Washington, D. C., to New York and on to Boston, it is next to impossible to imagine what our East Coast was like when the glaciers were here. The places where all the major cities are today were *under* the ice. The ice was so heavy that it pushed the land down below sea level, and the ice extended out into the ocean! That means we would need a boat to go north along the coast unless we wanted to risk walking over glaciers. It would have to be a big boat, because there were some awesome creatures in the ocean at that time—and they were probably hungry. If we were to get close to the edge of the ice sheet, we would have to be careful because every few minutes huge icebergs broke free and crashed into the ocean. (See Figure III-6-1.)

After many thousands of years, the climate got warmer and the glaciers and ice sheets melted. Tons and tons of mud, boulders, and sand were dumped everywhere by rushing rivers swollen with frothy gray meltwater. This muddy mess covered most of New England, and buried in it were the rounded boulders, slabs, and cobbles for Grampa's stone wall.

Just a little over 100 years ago, a young Swiss geologist, Louis Agassiz (ag-a-zee), suggested the idea of a time when ice thousands of feet thick covered large areas of North America and Europe. The reaction of learned men and women of the day was shock and amazement. But Agassiz had done his work carefully, and he had lots of data to support his story. He and his graduate students had measured the movement of glaciers in his native country of Switzerland, and he had spent several years in New England looking at tumbled debris that he realized resembled glacial sediments in the Alps. Agassiz found rocky outcrops and offshore islands in New England that had been rounded by the grinding pressure of huge rocks caught in the ice. Some of these rocky hummocks had shallow grooves and scratches on them, and it wasn't long before he figured out that the glaciers had left these marks. Scientists today call them glacial striations and use them to figure out which way the glacier was moving. Grampa found a piece of one of these scratched rocks and hid it in his wall.

Since Louis Agassiz, we have learned that the ice came several times and it will probably come back again some day. Field geologists in New England, especially on Nantucket Island, have found evidence for more than one glacial advance; and hardy scientists are analyzing ice in Antarctica and Greenland to learn more about very old ice sheets left from many glaciations. The scientists have drilled into ice nearly two

Figure III-6-1. The terminus of the Columbia Glacier, Prince William Sound, Alaska, in 1976. Where a glacier flows into the ocean, and icebergs are born.
Source: J. Schlee, Our Changing Continent, *p. 6-7, 1991. General Interest Publication, Department of the Interior/Geological Survey.*

miles thick, and they are analyzing layers that record annual events back to 160,000 years ago. They have found that the glaciations were about 10,000 to 20,000 years apart. It has been about 10,000 years since the last glaciation, so scientists expect we will have another one within the next 10,000 years or so unless other climatic changes occur to modify this cycle.

An ice advance happens when ice sheets and valley glaciers grow faster than they melt. The amount of snow affects how fast they grow, and the temperature during the summer affects how fast they melt. Some scientists think that slight changes in the earth's orbit around the sun or the tilt in the earth's axis can cause the right climatic conditions for an ice advance.

Geologically speaking, these glacial visitors are relative newcomers. They started arriving about 2 million years ago. Geologists call this period of time the Pleistocene epoch, and it has been referred to as the Great Ice Age. Glaciers have come and gone many times during the Pleistocene! With each glaciation, ice builds up in the northern and southern regions of our planet. The climate changes in the central part of the globe as well, even though there has never been ice near the equator. The warm times between glaciations are called interglacial periods, and we are living in one now.

"When will we have another ice age?" might be a better question to ask, and it is an important question, especially for those who design and build the docks for big ships in our coastal cities. There is a direct link between the amount of ice on the earth and sea level; during a glaciation, sea level drops as glacial ice builds up on the continents. During an interglacial period, like the one we are in now, sea level rises as the glaciers melt.

Scientists are watching to see what the glaciers are doing. Each year they carefully measure glaciers around the world and the ice sheets in Greenland and Antarctica to see if they are growing or shrinking. Over time, the change in size of the present glacial ice is an important indicator of modern global climate changes. And there is evidence that some of today's glaciers are still melting—but gathering good data from all over the world is a very big job. Scientists in many countries will have to work together, and the measurements will have to continue for several more years before we can be certain whether ice is increasing or decreasing.

Were humans here when there were lots of glaciers on the earth? Yes, they were. As a matter of fact, very new information has led some scientists to believe that it was *because* of the Ice Age that modern man got his start! The story of man's origins keeps changing as new fossil evidence is uncovered, new materials are dated, and old hypotheses are challenged. According to present ideas, about 3.5 million years ago Africa began to get drier and cooler with the onset of the Ice Age. An ancient primate, *Australopithecus* ("Lucy" is an Australopithecine fossil, about 3.2 million years old) had trouble finding food and a place to sleep because the trees and woodlands it inhabited were dying in the dry climate. It is not clear why, but about 2.5 million years ago, a small group of Australopithecines developed larger brains—much larger—and they figured out how to survive without climbing into the trees (which were being replaced by grassland). This new group survived, but the Australopithecines seem to have disappeared from the earth before 2.4 million years ago.

The new group, whose fossil remains date from about 2.4 million years ago, were our ancestors, *Homo rudolfensis*, the earliest species of the human genus *Homo*. They were different from their forebears in several important ways. The shape of their heads provided for expanded space in the front of their brains, allowing the ability to plan for the future and pass knowledge on to the next generation. Mothers did not need to climb into trees and had two free arms to care for their more helpless infants. The larger brains also meant that their children required longer to grow up and had to be cared for by two parents. So the males cooperated and hunted as a group instead of fighting. *Homo rudolfensis* had a language, learned to make tools, and found a way to make their own fire.

After nearly a million years of roaming the grasslands of Africa, *Homo rudolfensis* was replaced by a younger human species, *Homo erectus*. The oldest remains of *H. Erectus*, called "Turkana boy," were preserved 1.6 million years ago on an ancient floodplain in Kenya. This nearly complete skeleton suggests an evolution to an even larger brain size. *H. erectus* was the first species of the human family to make its way out of Africa. In fact, groups of *H. erectus* walked to Europe where the ice was. They were able to survive the ice because they learned how to make clothes from fur, made fire, and made distinctive stone tools called Acheulean. They were well organized, good hunters, and ended up being world travelers. *Homo erectus* items such as bones, fire pits where they cooked dinner, and fur clothing have been found in China, Indonesia, and Europe.

For more than a million and a half years, groups of *Homo erectus* ranged far and wide, apparently without much change in bone structure or brain size. But eventually in Europe, while glaciers still covered the landscape, a new individual appeared with an even larger brain; he was called *Homo neanderthalensis*. This fellow was apparently well established by about 300,000 years ago and was able to survive right in the middle of the coldest part of Europe during major glacial advances. His survival, scientists suggest, was because his body, like that of modern Eskimos, was stocky to conserve heat, and his nose cavity was large, which may have served well for warming and moistening the frigid, dry air of Ice Age winters. He had a larger brain than we do today, possibly in order to operate the heavier muscles in his robust body. There is uncertainty about whether *H. neanderthalensis* could speak. Some researchers suggest that his voice box wasn't in the right place for speech. The Neanderthals knew how to make their own fire, and they cared for their old and sick. They left a good fossil record because they carefully buried their dead. These glacial dwellers made better tools than their ancestor *Homo erectus*, but there was little change in the design of their implements over tens of thousands of years, suggesting to some scientists that they were not very innovative. An interesting fictional book portraying what Neanderthals may have been like is Jean M. Auel's *Clan of the Cave Bear*, which contains ideal passages to read aloud to your students.

Now what about us—*Homo sapiens*—tall, long-legged, with a high forehead, flat face, and prominent, willful chin? Interestingly, modern genetics supports fossil evidence that *Homo sapiens* evolved from a group of *Homo erectus* that remained in Africa. Genetic diversity in blood proteins and enzymes from African people today indicate that they have the longest history compared to European and Asian races. Scientists aren't sure exactly when our earliest ancestors appeared in Africa, but it was more than 100,000 years ago, long before Neanderthals had disappeared from Europe. The earliest remains of *H. sapiens* were found in what is now Israel, along with bones of Neanderthal of the same age! Perhaps the two species lived in harmony.

About 34,000 years ago *Homo sapiens* migrated northward to Europe and for a few thousand years may have competed with the Neanderthals for food, hunting grounds, or cave shelters in the midst of the most recent glacial advance. Neanderthals disappeared about 30,000 years ago, possibly as a result of competition from this quick-footed newcomer.

Activity I

How Do Glaciers Move ?

By Gary Vorwald, Island Trees High School, Levittown, NY, 1993.

OBJECTIVE: To give students the opportunity to observe and measure the movement of a glacier model.

> National Science Education Content Standard(s), K-4, 5-8
> Major Emphasis: A, C, E; Minor Emphasis: B;
> Other Content: Mathematics, Art, Language Arts

MATERIALS NEEDED:
Elmer's glue, borax solution, talcum powder, water, graduated cylinder, spoon, cup, water-resistant or permanent marker (e.g., Sharpie), food coloring, plastic tray or cookie sheet, metric ruler, stopwatch.

PROCEDURE:
1. Use the directions below, "How to Make Gak," to prepare your model glacier.
2. Place Gak on a tilted surface (cookie sheet or plastic tray with a book under one end) to represent a valley glacier.
3. Place another sample of Gak on a flat surface to represent continental glacier.
4. Using the marker, mark four or five points across the valley glacier. Alternatively use small buttons, paper clips, or push pins to mark starting points on the glacier. Place a line on the tray even with the points to indicate the starting point. Record starting time.
5. On the continental glacier place an X at the center, and mark four points around the perimeter of the glacier. Record starting time.
6. Observe the Gak glaciers at five-minute intervals. Measure the distance that the points have moved. Note any other observations about the glaciers.
7. Sketch the glaciers and points at the end of the observation period.
8. Calculate the rate of movement from the beginning of the investigation.
9. Write a conclusion about the rate that glaciers move. Note any differences in how valley glaciers differ in their movement from continental glaciers.

HOW TO MAKE GAK:
Safety Precautions: Gak is a product of the Mattel Toy Corporation that can be duplicated using common materials. Gak can be stretched, squeezed, twisted, and squished. Borax (sodium borate) is moderately toxic in quantities of more than one gram per 1,000 g of body weight. Wash any borax from the hands with water. Wash hands after handling the slime. Clean up any spilled Gak immediately.
1. Measure 5 ml (1 level teaspoon) of talcum powder (banned in some places—call your drug store for a substitute) and place it in a 5-oz paper cup. Add 25 ml (5 teaspoons) Elmer's glue, and 20 ml (4 teaspoons) water. Stir well to mix all ingredients.
2. Add up to 5 drops of food coloring to the materials in the paper cup and stir well.
3. Add 5 ml (1 teaspoon) of borax solution (1 level tablespoon of borax for each cup of water) and stir well.
4. Remove the Gak from the cup. Pull the solid off the stirrer. The Gak may be sticky at first but will become less sticky after handling. Dispose of the paper cup and any remaining liquid in the trash.
   *Recipe by Dave Katz, Flinn Scientific Catalogue, 1994.

Note: The Gak should stretch and flow easily, but it will tear if pulled hard. The Gak will dry out and become less stretchy after handling. It can be re-hydrated by mixing with a small amount of water before storage. Store the Gak in a plastic bag.

ACTIVITY II

What Happens Under a Glacier?

OBJECTIVE: To demonstrate that the melting temperature of ice decreases as pressure increases. Water melts under a glacier, providing a slippery layer for it to move on.

> National Science Education Content Standard(s), K-4, 5-8
> Major Emphasis: A, C, E; Minor Emphasis: B;
> Other Content: Measurement

MATERIALS NEEDED:
A smooth surface to make an inclined plane about 3 feet long and 1 foot wide, 8 inches high at one end.
Several ice cubes frozen to dryness—not damp. Take them out of the freezer just before the activity.
A variety of lead fish weights, nuts and/or bolts, buckshot, or BBs.

PROCEDURE:
1. Put the weights into the ice cube tray, add water, and freeze. Turn the cubes over and use the top of the cube for the activity. Divide the weights into four groups from heaviest to lightest, e.g., 1 oz., 2 oz., 3 oz., 4 oz., and 8 oz. The heaviest weight should be no more than half a pound
2. Line the ice cubes up across the inclined plane and see which one moves the fastest!

Tip: You may need to adjust the angle of the inclined plane.

RESULT: The heaviest ice cube should move down the inclined plane first.

ACTIVITY III

How Does a Glacier Make Striations or Glacial Scratches?

OBJECTIVE: To demonstrate that glacial scratches were made when harder rocks pressed into softer ones.

> National Science Education Content Standard(s), K-4, 5-8
> Major Emphasis: A. C. E; Minor Emphasis: B;
> Other Content: Organization

MATERIALS NEEDED:
Moh's Scale of Mineral Hardness (available through most science catalogs)
Ten minerals of different hardness

PROCEDURE:
1. Have a group of students test each of the minerals and arrange them in order of hardness.
2. Use the harder minerals to make striations (scratches) on the softer minerals.
3. Put several of the harder rock chips in cups of water—freeze them, and then take them out of the container and scratch them across a cement sidewalk.

ACTIVITY IV

Where Are Modern Glaciers?

OBJECTIVE: To demonstrate that glaciers are found at high altitudes and in polar regions.

MATERIALS NEEDED:
A map of the world showing modern glaciers and ice sheets (a good atlas will have maps like these).

PROCEDURE:
1. Locate ten mountain glaciers on a map of the world and note for each:
   • The altitude above sea level;
   • The beginning, or head of the glacier, and its terminus, or end.
2. Locate two ice sheets on a map of the world and note for each:
   • The latitude (degrees above or below the Equator) at which they are found;
   • The continent upon which they are found.

RESULT: 1. Mountain glaciers occur at high *altitudes* at any latitude.
2. Ice sheets are found at high *latitudes*.

ACTIVITY V

How Heavy Is a Glacier?

OBJECTIVE: To demonstrate that glaciers are heavy, and that's why they can push land down below sea level.

EQUIPMENT NEEDED:
A plastic cylinder, sealed at one end, exactly one foot long, and 3" or 4" in diameter; water; a freezer; a scale to weigh the cylinder.

PROCEDURE:
Divide your students into groups and have each group do part of the activity, or, obtain enough cylinders to have each group do the activity.
Group 1: Weigh the empty cylinder very carefully.
Group 2: Nearly fill the cylinder with water to within *one inch of the top*.
Group 3: Freeze the cylinder upright—add water as necessary to bring the level of ice exactly to the top of the cylinder. Remember, ice expands!
Group 4: Weigh the frozen cylinder.
Now, use your data to figure out how much a glacier could weigh!
Whole class: Imagine your cylinder and many, many more like it stacked on top of each other—enough to make two miles of glacial ice (that was about the maximum thickness of ice sheets in North America). Multiply the weight of your cylinder times the number of feet (cylinders) in two miles (5,280 feet in one mile times 2). It will be a big number.

QUESTIONS FOR STUDENT RESEARCH:
• How far down did the glaciers push the land in New England? [The land in some places was pushed down as much as 500 feet, and has come back up. This is called glacial rebound.]
• How do we know when the climate changes? Who measures the change?
• Where do icebergs come from today?

REFERENCES AND FURTHER READING:

Crawford, Michael, 1998. ORIGINS OF NATIVE AMERICANS: Cambridge University Press.

Schlee, John, 1991. OUR CHANGING CONTINENT: U.S. Dept. of the Interior/U.S. Geological Survey, U.S. Government Printing Office, Washington, DC, 21 pp.

Stanley, Steven M., 1998. CHILDREN OF THE ICE AGE: W. H. Freeman and Company, New York, 278 pp.

USEFUL WEBSITES:

Origins of Humankind
•http://www.dealsonline.com/origins/research/habilis.htm
Site maintained by Jeff Rix, Wexford, PA.

Problem-Solving Activities
•http://www.ibm.pbscyberschool.org/index2.html
Math and science for middle school kids focused on downhill skiing, snow boarding, and figure skating.

Hominids
•http://lrs.ed.uiuc.edu/students/b-sklar/basic387.html
Excellent summaries of hominid species.

Humans and Glaciers
•http://www.museum.state.il.us/exhibits/larson/overkill.html
Relationships between Homo sapiens, ice retreat, and extinct fauna. Terrific animation of ice retreat 18,000-7,000 years ago.

Ice Age
•http://www.museum.state.il.us/exhibits/larson/content.html
Midwest U.S., 16, 000 years ago, from the Illinois State Museum.

Paper Models for Students to Make
•http://www.usgs.gov/education/animation/ice89-/90-/90sea.hqx
Models show the effects of glacial ice on a mountain valley.

III-7

What Happened to the Woolly Mammoths?

Grampa cleared his throat. "Then, out of the dust sprang a spongy green carpet of plants called tundra. Tiny flowers and miniature shrubs hugged the ground in that cooler, windier world. Caribou grazed in great herds. Woolly brown mammoths trampled the tundra, waving their trunks in the air."

From *Stone Wall Secrets*

OBJECTIVE: To explore the extinction of a species and the factors that might have caused the mammoth to disappear.

What was this humongous, furry animal with the giant curved tusks that Grampa "heard" rumbling and trumpeting across the tundra? Was it an elephant with fur? No, it was a separate species. About three million years ago, long before the beginning of the Great Ice Age, the woolly mammoth and the elephant probably evolved from the same common ancestor. The woolly mammoth is closely related to one of today's elephants—the one that is found in India, the genus *Elephas*. But the mammoth had fur—in fact, some of his fur hung down more than two feet.

Perhaps it was this long, thick fur that allowed the mammoth to survive the glacial climate, because he apparently lasted for two million years through all the glacial stages of the Pleistocene. In between advances of the glaciers, our woolly friend walked to every continent but Antarctica. He migrated, like our ancestor

Homo sapiens, across the Bering Land Bridge into Alaska and down the length of North and South America. As recently as 12,000 years ago, herds of mammoths may have been as common as white-tailed deer are today! Grampa "heard" them rumble into New England right behind the retreating ice.

But Grampa and Adam and the rest of us will never really hear or see a woolly mammoth. They are all gone! Why? What happened to them? The elephant is still here—possibly because during the many Pleistocene glaciations, he was hiding out further south, beyond the icy clutch of the glaciers.

The last woolly mammoth died about 2,000 years ago, before humans kept records of endangered species, but we do have pictures of the beasts left on cave walls. In France 25,000 years ago, our talented ancestors left beautiful drawings of the woolly mammoth. And in Siberia, early humans built huts from mammoth bones. Does this mean that humans hunted these beasts to extinction? Scientists argue about this—some are sure that as the numbers of *Homo sapiens* increased and his talent for hunting got more proficient, the mammoth didn't have enough young to support a stable population and they gradually dwindled and then were gone.

IDEA FOR AN ACTIVITY:
• Research the differences between elephants and woolly mammoths.

REFERENCES AND FURTHER READING:
Ward, Peter, 1997. THE CALL OF DISTANT MAMMOTHS, WHY THE ICE AGE MAMMALS DISAPPEARED: Copernicus, Springer-Verlag, New York, NY, 241 pp.

USEFUL WEBSITES:

NOVA Site for Teachers Guides
•http://www.pbs.org/wgbh/nova/teachersguide/mammoth/
•http://www.pbs.org/wgbh/nova/teachers/

Prehistoric Culture in Alaska
•http://www.blm.gov/education/mesas/clues.html
A science detective story about a possible new prehistoric culture in Alaska—with some good descriptions of how carbon-14 dating is used.

Great Mammoth Resource Site
•http://www-cot.idbsu.edu/~btech/bsuradio/misc/mammoth/toc.html
For teachers. Created in association with the Idaho State Historical Society.

Radiocarbon Dating
•http://packrat.aml.arizona.edu/Journal/v37n1/vartanyan.html
Evidence for mammoths on Wrangel Island, Arctic Ocean, until 2000 BC. This is a good example of a scientific article and speaks to the debate about what caused mammoth extinction.

Ice Age Paleoecology:
•http://culter.colorado.edu:1030/~saelias/elias.html
The home page for Scott Elias at the Institute of Arctic and Alpine Research, University of Colorado, Kineo

III-8
Rocks for Tools

OBJECTIVES: To show how early inhabitants of North America used minerals and rocks; and to show how archaeological detectives are using these items to understand early cultures.

I can imagine Adam asking how many years ago the Indians sat beside the campfire that left its telltale red oxide on the hearthstone from Grampa's wall. How would Grampa answer? Well, Grampa might turn the stone over to expose the red, burned side, and maybe he would find a black smudge of charcoal (carbon) left on the stone from the fire pit. He would point to the black smudge and explain to Adam that it was a radioactive clock! (See, "How Do We Measure Time?" Section I-5.) The amount of carbon-14 left in the charcoal would tell us the age of the firewood burned next to the hearthstone.

Could we tell how old the fire pit was by the age of the material on the inside of the hearthstone? No, because many of the rocks in New England are hundreds of millions of years old, and we know the first people on this land are much, much younger than that.

The earliest people to sit around campfires in New England might have followed the woolly mammoths and other tundra animals across the Bering Land Bridge to Alaska and then south about 12,000 years ago. They brought the knowledge of how to make fire with them.

How can scientists tell what materials the Indians used to make their pots? From a single broken piece of pottery (called a "pot shard"), archaeological detectives can tell that the Indians used clay to make the pot, and where in New England the clay came from. They analyze the kinds of minerals and chemical elements in the pot shard with microscopes and very smart machines that perform chemical analyses. Only certain kinds of clay are useful for pots, and each area where clay is found has a chemical signature—that is, a group of chemical elements that are common to that area.

Indians make their pots from the same clay that was used to make bricks for early farmhouse chimneys! The clay is called "blue clay" and it was kindly left behind by the glaciers. The glaciers ground up rocks into what earth scientists call "rock flour." Tons and tons of this fine material washed from under the glaciers and into the shallow sea that covered the New England coast 20,000 years ago. Today, layers of blue clay are very common, especially along coastal areas of Maine.

Certain clever Indians figured out that if they added shells to their clay, the pots would be much stronger. The shells were ground up fine and mixed with the pot clay. When the pot was fired (heated very hot), the shell material changed the clay into a mineral that helped to make the pot stronger. When pieces of these pots are found, archaeologists know what kind of Indians made them.

Shells were used for money, or *wampum*—so coastal areas were popular places for the Indians to come and eat oysters and trade their goods for *wampum*. Today we find many mounds of shells that archaeologists tell us were ancient garbage heaps. Because the mounds are on the coast, some were washed away or submerged by sea level changes. Remember, the sea level came up when the glaciers melted, but in some areas, it also fell as the land came up like a cork after the glaciers were gone. The early Indians had to deal with these changes.

About 5,000 years ago a group of New England Indians called the Red Paint People used a variety of rock materials in some interesting ways. They ground hematite (iron oxide) into a red powder, which they sprinkled over their dead before burial. They made fire by striking flintstone and pyrite (iron sulfide, FeS_2). A sharp blow would cause tiny pieces of pyrite to make big sparks that lasted long enough to ignite dry leaves. They carried the fire-making kits with them in a pouch. We can imagine these early natives trading goods for flintstone, which was so necessary for fire. Slate and quartzite from the shores of the ancient Iapetus Sea were used for crafting tools and ornaments.

When the first European settlers came to New England in the middle 1600s, the Indians had a thriving trade among groups in different areas, and they apparently had been trading over long distances for thousands of years. Throughout North and South America, the crushing of tectonic plates created many kinds of rock the Indians could use and trade. Many are shown in Figure III-8-1. Serpentine, or soapstone (from the ocean floor munched up 400 million years ago along the northeast coast of New England) was carved into ornaments, and talc was used to make things slide more easily (like talcum powder is used today). Indians from the Northeast may have gone all the way to South America to get jade, which was far more precious than gold, and black, volcanic obsidian, which made the sharpest knives. Traders carrying copper and flint came down from Nova Scotia, and there were regular runs made to Mt. Kineo on Moosehead Lake in Maine for Kineo felsite. Felsite is a very, very fine-grained rock made of quartz and feldspar. It made sharp, hard arrowheads and spear points. The first Europeans may have had to duck as these Kineo felsite arrows came flying at them!

CHOICE OF STONES

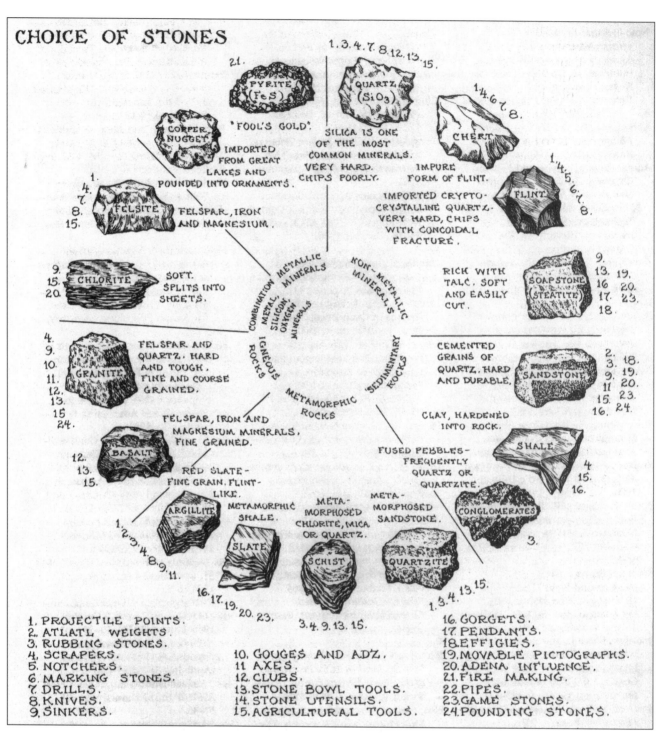

1. PROJECTILE POINTS.
2. ATLATL WEIGHTS.
3. RUBBING STONES.
4. SCRAPERS.
5. NOTCHERS.
6. MARKING STONES.
7. DRILLS.
8. KNIVES.
9. SINKERS.
10. GOUGES AND ADZ.
11. AXES.
12. CLUBS.
13. STONE BOWL TOOLS.
14. STONE UTENSILS.
15. AGRICULTURAL TOOLS.
16. GORGETS.
17. PENDANTS.
18. EFFIGIES.
19. MOVABLE PICTOGRAPHS.
20. ADENA INFLUENCE.
21. FIRE MAKING.
22. PIPES.
23. GAME STONES.
24. POUNDING STONES.

Figure III-8-1. Stones in Mother Nature's General Store. Early Indians survived by using rocks and minerals to hunt, fish, cook, socialize, and play. A thriving trade kept Indian groups happy and well fed.
Source: C. Keith Wilbur, The New England Indians, *Copyright 1996, Globe Pequot Press.*

- Find out about Indian trade routes in New England or your area.
- Form two tribes and trade for the things you need. For example, corn for chert (arrowheads)
- Invite a local Native American to your class to talk about how his ancestors lived.
- Visit a museum and research where early people lived in your area. Draw pictures or a mural showing how they lived.

QUESTIONS FOR FURTHER STUDENT RESEARCH:

How did early humans keep warm during the Great Ice Age?
Did humans contribute to the extinction of the woolly mammoth?

REFERENCES AND FURTHER READING:

Braun, Esther, and David Braun. THE FIRST PEOPLES OF THE NORTHEAST: Lincoln Historical Society, Lincoln, MA.
Wilber, C. Keith, 1996. THE NEW ENGLAND INDIANS: Globe Pequot Press, Old Saybrook, CT, 88 pp.
Wilber, C. Keith, 1995. THE WOODLAND INDIANS: Globe Pequot Press, Old Saybrook, CT, 128 pp.

USEFUL WEBSITES:

ArchNet from the UConn Library
- http://spirit.lib.uconn.edu/ArchNet/ArchNet.html
Links to sites with archeology information.

Archaeology Magazine
- http://www.he.net/archaeol/index.html
This page is linked to the AIA home page which has links to good resources for teachers.

Center for the Study of the First Americans
- http://www.peak.org/csfa/csfa.html
Lots of information on early tool-making and fluted points. The link to the Encarta overview of the First Americans is excellent and current.

People and Environmental Change on the Northern Great Plains
- http://www.usd.edu/anth/epa
From the University of South Dakota, there's a link to a page called "The First Peoples, 10,000 BC—Did Overhunting Cause the Mammoth to Become Extinct?"

Native American Archaeology and Anthropology Resources on the Internet
- http://hanksville.phast.umass.edu/misc/indices/NAarch.html
This site has links to pages with Native American archaeology and anthropology information.

III-9

Ecology:
Action/Reaction and Balance

OBJECTIVE: To show that living things are constantly seeking a balance as conditions change.

> *These colonists from Europe changed the wilderness from forest into farm. They chopped down acre after acre of tall timber, burning the trees in bonfires that glowed orange and red through long days and nights. Oxen bellowed and clanked their chains as they tore the stumps from the ground. Man and beast worked as a team to change the world forever.*
>
> From *Stone Wall Secrets*

Have you ever noticed a vaccination scar on someone's arm? Or gotten together with your friends and compared battle scars? The time you cut your leg when you fell out of the treehouse, or when you got that splinter in your finger? The scar is your skin's reaction to change—to being cut or scraped. Forests also heal when they are cut. And like your skin, the wounded area may never be exactly like it was before.

Will Grampa's fields ever return to the way they were when the first settlers got here? Scientists are trying to answer this question, and they say there hasn't been enough time to tell yet, but even after so many years, there are big differences between plowed fields and forests that have never been cleared.

Let's look at what is different about Grampa's field today compared to when the settlers worked so hard to clear their land. The trees today are much younger and therefore are much smaller, and there is much more sunlight reaching the ground. That means sun-loving plants would be favored over those needing shade. And look at the big differences in the soil. Gone is the spongy black humus layer of decaying leaf litter and animal droppings. Today there is only a thin layer of muddy soil with a healthy number of cobbles and boulders in it. Many muddy springtimes have come and gone since the oxen were hitched up and Grampa helped Uncle Cyrus move stones out of the field. Only certain types of seeds can work their way down into this kind of soil, and they are different than the ones that 300 years ago found their way into the soft spongy forest litter.

Perhaps you can begin to see that plants today in woodlands and fields surrounded by the old stone walls are in the process of changing. These fields are an ecosystem, now characterized by young trees and shrubs. When the settlers arrived, the ecosystem was forest, and when humans cleared the land they turned it into grassland. Today the grassland is in the process of returning to forest. This process is a bumpy one and can quickly change direction every time we have several cold winters in a row, or several dry summers, and plants die away to be replaced by others more able to stand the weather conditions. It was exactly the same process that brought the forests back after the glaciers melted. Grampa describes the changes in the land:

> *"The tundra receded northward when the climate began to warm. The animals migrated, too, leaving bleached antlers and tusks behind. The trees returned—first spruce and pine, next birch and oak, then finally, all the rest. Their roots cracked the hard, glacial soil, breaking it apart. Dust and leaves mixed together with stones and sand to make a rich, dark soil. The forest filled with woodland creatures—wolves, bears, panthers, possums, and deer."*
>
> From *Stone Wall Secrets*

Does this process of adjusting to change ever stop? To answer that we would have to ask another question. Do conditions ever stop changing? Well, we can be pretty sure that climate change is always occurring. We have seen how glaciers have come and gone. And we know that longer term changes are occurring because the tectonic plates are constantly driven across the surface of the earth, from one climatic region to another. As continents move closer to the equator they get warmer, and toward the poles, like the position of Antarctica today, the climate is much colder. These changes are very slow—we don't even know they are happening most of the time. But our scientist friends are very glad when early settlers kept records of the kinds of trees that were growing on their lands and what animals lived among them—because things have changed so much in just a few hundred years that we can't be sure what that earlier, shady ecosystem was like.

ACTIVITY I

Grow the Same Plants in Soils Made from Different Crushed Rocks

OBJECTIVE: To demonstrate that rock type is an important factor in the kind of plants that grow in an area .

National Science Education Content Standard(s), K-4, 5-8
Major Emphasis: A, D, G; Minor Emphasis: B;
Other Content: Measurement

MATERIALS NEEDED:
- Approximately one cup each of crushed quartz sandstone, granite, limestone, and volcanic rock. Most

of these can be obtained from your state Geological Survey office, or purchased from educational supply companies.
- At least 25 seeds of the same kind of plant (any plant seeds can be used—but it would be good to have information on what type of soil is preferred by the plant you choose).
- Soil pH test kit from a garden store.
- One cup of potting soil (available from a garden store).
- Five plastic flower pot containers with drainage hole and five saucers to put under the pots to hold drainage.

PROCEDURE:
1. Divide your class into five groups and give each group one of the crushed rock samples (have them draw from a hat to determine who gets what).
2. Place crushed rock samples in plastic containers. Leave about 1/2 inch at the top of the container to put in water. Put the very finest material at the top of the pot to plant the seeds in. Set aside two heaping teaspoons of this fine material. Place potting soil in a separate container. It will be your control.
3. Determine the pH of the potting soil and of your crushed material using the two teaspoons of fine material set aside in Step 2. Follow directions on soil pH test kit. (See Note about soil pH below.)
4. Plant at least 5 seeds in each pot.
5. Instruct each group to assign one member to water the plant seeds every day until they sprout. (Follow the directions on the seed packet about having the seeds sprout.)
6. After the seeds have sprouted only water them when the "soil" is dry (about every other day).
7. Observe the results at the end of each week. Have the groups each measure the growth of their plants. (Have a class discussion on what to measure.)
8. Make sure the groups put their measurement data on a table. If they don't measure one week, be sure they report "no data" that week, etc.
9. Graph the results as growth (in some unit of length) against time (horizontal axis).
   Tip: If you use cut pieces of potato with an "eye," you may be able to sprout quicker; potatoes are very sensitive to pH of the soil (they prefer acid soil).

Note: Soil pH is an indication of the acidity or alkalinity of soil and is measured in pH units. Soil pH is defined as the negative logarithm of the hydrogen ion concentration. The pH scale goes from 0 to 14 with pH 7 as the neutral point. As hydrogen ions in the soil increase, the soil pH decreases thus becoming more acidic. From pH 7 to 0 the soil is increasingly more acidic and from pH 7 to 14 the soil is increasingly more alkaline or basic.

DESCRIPTIVE TERMS COMMONLY ASSOCIATED WITH CERTAIN RANGES IN SOIL pH ARE:

- extremely acid, < than 4.5; lemon = 2.5; vinegar = 3.0; stomach acid = 2.0; soda = 2 - 4
- very strongly acid, 4.5 - 5.0; beer = 4.5 - 5.0; tomatoes = 4.5
- strongly acid, 5.1 - 5.5; carrots = 5.0; asparagus = 5.5; boric acid = 5.2; cabbage = 5.3
- moderately acid, 5.6 - 6.0; potatoes = 5.6
- slightly acid, 6.1 - 6.5; salmon = 6.2; cow's milk = 6.5
- neutral, 6.6 - 7.3; saliva = 6.6 - 7.3; blood = 7.3; shrimp = 7.0
- slightly alkaline, 7.4 - 7.8; eggs = 7.6 - 7.8
- moderately alkaline, 7.9 - 8.4; sea water = 8.2; sodium bicarbonate = 8.4
- strongly alkaline, 8.5 - 9.0; borax = 9.0
- very strongly alkaline, > than 9.1; milk of magnesia = 10.5; ammonia = 11.1; lime = 12.

Source: Donald Bickelhaupt, Faculty of Forestry, State University of New York, College of Environmental Science and Forestry, 1993.

Activity II

Locate, Photograph, and Name Local Ecosystems

OBJECTIVE: To identify ecosystems and their associated soils.

> National Science Education Content Standard(s), K-4, 5-8
> Major Emphasis: A, D, G; Minor Emphasis: B;
> Other Content: Art

An ecosystem *is a community of organisms and its nonliving, physical environment. For example: an abandoned field with stone walls around it.*

MATERIALS NEEDED:
Disposable cameras—one for every five students (Suggestion: Have your parent-teachers organization fund this class project. As incentive, the results can be given to the local newspaper!)

PROCEDURE:
1. Using disposable cameras, have each group of students take pictures of as many local ecosystems as possible. Use all the pictures in the camera and get several that will help to define the ecosystem. (Suggestion: Divide your local area into sections and form your class groups from students who live in the sections. For urban areas, ask for a parent volunteer to go with each group of students.) Be sure the group takes notes on each area photographed, and on each photograph.
2. Develop the pictures and have a class discussion during which each group describes the ecosystems they photographed.
3. Use the photographs to identify characteristics that are common to each ecosystem, and things that are different, if any.
4. Using your list of characteristics, identify and give names to the major ecosystems.
5. Make a list of all ecosystems identified, and be sure the class copies the list.
6. Compare major categories with a soil map of your area from the U.S. Department of Agriculture Natural Resources Conservation Service (formerly Soil Conservation Service).

Activity III

Observe Changes in Local Habitats within Ecosystems

OBJECTIVE: To identify and study habitats within ecosystems and learn the difference between them .

> National Science Education Content Standard(s), K-4, 5-8
> Major Emphasis: A. D. G; Minor Emphasis: B;
> Other Content: Measurement

A habitat *is the natural environment or place where an organism, population, or species lives. For example: chipmunks sometimes live in a stone wall.*

MATERIALS NEEDED:
Clipboards; notebooks; tape measures; magnifying glasses; binoculars

PROCEDURE:
1. Using the photographs of the ecosystems from Activity II, have a class discussion about what habitats might be found in each ecosystem.
2. Have each group of students return to their systems and choose one habitat from one ecosystem to observe for the next several months. (Before making the final selections have the groups get back

together and name their choices of habitats. Make sure that the groups have not chosen the same ones. And make sure the habitat they have identified is safe and practical for them to visit regularly.)

3. For each habitat, have the group make a map, a drawing, and a list of things to observe on each visit.

QUESTIONS FOR STUDENT RESEARCH:
- What kind of trees grow in soil from sandstone? Granite? Limestone? Glacial debris? Volcanic rocks?
- What is tree pollen?
- What can pollen studies tell us about major events like the process of cutting down trees to clear land?
- What kind of moss grows on stone walls?
- What are lichens? What can lichens tell us about air quality?
- What kind of animals live in stone walls?

REFERENCES AND FURTHER READING:
Carson, Rachel, 1992. UNDER THE SEA WIND: Truman Talley Books/Plume Printing, New York, NY, 304 pp.
Watt, Fiona, 1991. PLANET EARTH, A PRACTICAL INTRODUCTION WITH PROJECTS AND ACTIVITIES: Usborne Publishing Ltd., London, UK, 48 pp.

USEFUL WEBSITES:

Satellite Images
•http://edcwww.cr.usgs.gov/Earthshots/
Earthshots is a collection of satellite images that show how the environment has changed over the last 20 years. The effects of droughts, fires, deforestation, urbanization, irrigation, desertification, and other natural events can be detected, measured, and analyzed using the data.

The Earth Balloon
•http://www.earthballoon.com/
The Earth Balloon is a traveling outreach program that features a 22-foot (6.7 meters) globe students enter to learn about the changing earth. Inside the Earth Balloon students learn about pollution, environmental science, earth science, geography, and math. The program is designed for grades K-12 with a packet of pre-visit materials being sent 2 weeks before a scheduled visit. The Earth Balloon is also available for purchase as a complete module. For more information, contact: david.knutson@earthballoon.com

World Builders
•http://curriculum.calstatela.edu/courses/builders/
This is an interesting site that is based on a course out of Cal Tech where students and/or teachers design planets that support a variety of life forms. It includes lessons, many links to web pages, teacher resources, science notes, and planets created by teams of teachers who participated in the course. It is being tried out in a high school, and bits of it have been used in the elementary grades. The website is in a state of ongoing construction, but over a hundred and fifty professor-written pages are already in place.

Ecology Field Studies for Educators
•http://maine.maine.edu/~eaglhill
Humboldt Field Research Institute, Steuben, ME, offers field experiences in five different ecological communities: lake, bog, salt marsh, stream, and red spruce-fir forest; formation, hydrology, soils, physiochemical relationships, and biota of these communities; student-centered learning approach, field experiences, study guides to help teachers to develop or expand their own field experience-based curriculum. Instructors: Dr. Herman Weller, University of Maine and Michael D. Warren, Monmouth Academy. Graduate credit available. For additional information on this course and other courses and seminars offered through Humboldt, call 207-546-2821 or email <humboldt@nemaine.com>.

Learning about Wetlands.
•http://www.athena.ivv.nasa.gov
An Athena site.

APPENDIX 1

Glossary

If no reference is given, the definitions were taken from the web site, Webster On-Line
http://www.m-w.com/dictionary
or from the AGI Glossary of Geology, 1987, 3rd Ed., Bates and Jackson, American Geological Institute, Alexandria, VA.

Acid	— a sour substance — any of various typically water-soluble and sour compounds that in solution are capable of reacting with a base to form a salt, redden litmus, and have a pH less than 7
Alkaline	— having the properties of an alkali or alkali metal — BASIC, especially of a solution having a pH of more than 7
Altitude	— the vertical elevation of an object above a surface (as sea level or land) of a planet or natural satellite
Archaic	the period from about 8000 B.C. to 1000 B.C. and the North American cultures of that time
Artificial Intelligence	branch of computer science which aims to imitate the thought processes of the human brain.
Asteroids	any of the small celestial bodies found especially between the orbits of Mars and Jupiter
Atom	the smallest part of an element that can exist; it consists of a nucleus of protons and neutrons surrounded by a cloud of orbiting electrons. (Couper and Henbest, 1997)
Avalonia	ancient landmass to the east of North America across the Iapetus Sea, 500 million years ago. Avalonia ultimately collided with North America 380 million years ago at the start of the Acadian Mountain Building cycle
Basalt	a dark-colored igneous rock that is often volcanic
Black Hole	a celestial object with a gravitational field so strong that light cannot escape from it which is believed to be created in the collapse of a very massive star
Breccia	a rock consisting of sharp fragments embedded in a fine-grained matrix (as sand or clay)
Cambrian	of, relating to, or being the earliest geologic period of the Paleozoic era or the corresponding system of rocks
Carbon	a nonmetallic, chiefly tetravalent (+4) element found native (as in the diamond and graphite) or as a constituent of coal, petroleum, and asphalt, of limestone and other carbonates, and of organic compounds or obtained artificially in varying degrees of purity especially as carbon black, lampblack, activated carbon, charcoal, and coke
Carbon Dioxide	a heavy colorless gas (CO_2 —one atom of the element carbon, and two atoms of the element oxygen) that does not support combustion, dissolves in water to form carbonic acid, is formed especially in animal respiration and in the decay or combustion of animal and vegetable matter, is absorbed from the air by plants in photosynthesis, and is used in the carbonation of beverages
Carbon Monoxide	a colorless odorless very toxic gas (CO) that burns to carbon dioxide with a blue flame and is formed as a product of the incomplete combustion of carbon
Carboniferous	— the geologic period of the Paleozoic era between the Devonian and the Permian or the corresponding system of rocks — includes extensive coal beds —see Chart of Major Events, Section I-2
Chemical Weathering	a process caused by the effects of weather in which chemical reactions transform rocks and minerals into new chemical combinations
Chicxulub	a location in Yucatan, Mexico, where large a large meteorite hit the earth 65 million years ago
Colleagues	partners who study together
Comets	a celestial body that consists of a fuzzy head usually surrounding a bright nucleus, that has a usually highly eccentric orbit, and that often, when in the part of its orbit near the sun, develops a long tail which points away from the sun
Conglomerate	rock composed of rounded fragments varying from small pebbles to large boulders in a fine-grained matrix of sand or silt, and commonly cemented by hardened clay, iron oxide or calcium carbonate
Continental Drift	a slow movement of the continents on a deep-seated viscous zone within the earth

Convergent Plate Boundary	a boundary between two tectonic plates that are moving toward each other
Cosmic	of or relating to the cosmos, the extraterrestrial vastness, or the universe in contrast to the earth alone
Cosmic Inflation	a very short period during the first second of the life of the universe in which it contained so much energy that it blew up and expanded trillions of times; postulated by Alan Guth of Stanford Linear Accelerator Center in 1979; inflation explains why the universe is so big and smooth, why different forces act in it today, and where the vast amount of matter came from
Cretaceous	— the last period of the Mesozoic era characterized by continued dominance of reptiles, emergent dominance of angiosperms, diversification of mammals, and the extinction of many types of organisms at the close of the period — the corresponding system of rocks —see Chart of Major Events, Section I-2
Devonian	— the period of the Paleozoic era between the Silurian and the Mississippian or the corresponding system of rocks — see Chart of Major Events, Section I-2
Ecosystem	the complex of a community of organisms and its environment functioning as an ecological unit in nature
Electron	a subatomic particle that has a negative charge (Couper and Henbest, 1997)
Element	a substance with unique physical or chemical properties that cannot be broken down into anything simpler by means of chemical reactions (Couper and Henbest, 1997)
Fossil	a remnant, impression, or trace of an organism of past geologic ages that has been preserved in the earth's crust
Frost Heave	an upthrust of ground or pavement caused by freezing and expansion of water in moist ground
Galactic	of or relating to a galaxy and especially the Milky Way galaxy
Gaseous Matter	equivalent to "cosmic dust"—particles of solid matter moving about in cosmic space such that their movements obey the behavior of a gas
Geologic time	the long period of time occupied by the earth's geologic history
Glaciation Glacier	to cover with a glacier, and to produce glacial effects in or on a large body of ice moving slowly down a slope or valley or spreading outward on a land surface
Granite	an igneous rock with visibly crystalline texture formed essentially of quartz, feldspar (orthoclase and/or plagioclase) and some dark minerals like biotite or hornblende; and used especially for building and for monuments
Habitat	— the place or environment where a plant or animal naturally or normally lives and grows
Halite	the mineral name for common table salt, NaCl or sodium chloride
Hazardous Waste	waste products of human activity that pose a present or potential danger to human beings or other organisms because it is toxic, flammable, radioactive, explosive, or has some other property that produces substantial risk to life
Helium	a light colorless nonflammable gaseous element (He) found especially in natural gases and used chiefly for inflating airships and balloons, for filling incandescent lamps, and for cryogenic research
Hot Spot	— a place in the upper mantle of the earth at which hot magma from the lower mantle upwells to melt through the crust usually in the interior of a tectonic plate to form a volcanic feature — a place in the crust overlying a hot spot
Hydrogen	a nonmetallic element (H) that is the simplest and lightest of the elements, is normally a colorless odorless highly flammable diatomic (two atoms, or H2) gas
Hydrogen Chloride	a colorless pungent poisonous gas (HCl) that fumes in moist air and yields hydrochloric acid when dissolved in water
Hypothesis	a tentative assumption made in order to draw out and test its logical or empirical consequences
Ice Core	a long cylinder of ice that slides up inside a metal tube like a gigantic soda straw as the metal tube is drilled into the ice
Iridium	a silver-white, hard, brittle, very heavy metallic element
Iron Oxide	any of several oxides of iron: hematite, magnetite, goethite, limonite, etc.
Juan de Fuca Plate	a tectonic plate that is being subducted under the west coast of the North American continent

Jurassic	the geologic period of the Mesozoic era between the Triassic and the Cretaceous or the corresponding system of rocks marked by the presence of dinosaurs and the first appearance of birds
Latitude	the angular distance from some specified circle or plane of reference; especially, the angular distance north or south from the earth's equator measured through 90 degrees
Leptons	any of a family of particles (as electrons, muons, and neutrinos) that have a spin quantum number $^{1}/_{2}$ and that experience no strong forces
Lithium	a soft silver-white element (Li) of the alkali metal group that is the lightest metal known and that is used in chemical synthesis and storage batteries
Longitude	— the angular distance measured on a great circle of reference (on the earth, this is the Equator) from the intersection of the adopted zero meridian (on earth, the Greenwich Meridian, through Greenwich, England) with this reference circle to the similar intersection of the meridian passing through the object — the arc or portion of the earth's equator intersected between the meridian of a given place and the prime meridian and expressed either in degrees or in time
Lunar Eclipse	the earth's shadow is cast on the moon when the earth is directly between the Sun and the Moon
Magma	molten rock material within the earth which rises into the crust and cools to produce igneous rocks
Magnetite	— a black, isometric, strongly magnetic mineral of the spinel group that is an oxide of iron and an important iron ore — it is often called lodestone
Mantle	— the part of the interior of the earth that lies beneath the lithosphere (or crust) and above the central core — the mantle is divided into the upper and lower mantle, with a transition zone between
Mechanical Weathering	the process by which frost action, salt-crystal growth, absorption of water, and other physical processes break down a rock to fragments, involving no chemical change.
Meteor	a transient fiery streak in the sky produced by a meteoroid passing through the earth's atmosphere
Meteorite	—a mass of stone or metal that has reached the earth from outer space —a meteor that reaches the surface of the earth without being completely vaporized
Meteoroid	a small body traveling through space
Mica	any of various colored or transparent mineral silicates crystallizing in monoclinic forms that readily separate into very thin leaves
Milky Way Galaxy	the galaxy of which the sun and the solar system are a part and which contains the myriad stars that comprise the Milky Way
Mineral	— a naturally occurring inorganic element or compound (more than one element) having an orderly internal structure (of atoms) and characteristic chemical composition, crystal form, and physical properties; those who include the requirement of crystalline form in the definition would consider an amorphous compound such as opal to be a mineraloid
Molecule	the smallest particle of a substance that retains all the properties of the substance and is composed of one or more atoms
Neutron	neutral particle, found only in the nucleus of an atom (Couper and Henbest, 1997)
Nitrogen	a colorless, tasteless, odorless element that as a diatomic gas (N_2) is relatively inert and constitutes 78 percent of the atmosphere by volume and that occurs as a constituent of all living tissues
Nucleus(plural = nuclei)	the positively charged central portion of an atom that comprises nearly all of the atomic mass and that consists of protons and neutrons except in hydrogen which consists of one proton only
Obsidian	an igneous volcanic rock, usually dark-colored, glassy, and characterized by conchoidal fractures leaving sharp edges so that it has been widely used for arrowheads
Ordovician	— the geologic period between the Cambrian and the Silurian or the corresponding system of rocks —see Chart of Major Events, Section I-2
Orogeny	the process of formation of mountains
Paleoindians	—see Chart of Major Events, Section I-2

Paleontology	a science dealing with the life of past geological periods as known from fossil remains; a paleontologist is one who studies paleontology
Paleozoic	— an era of geological history that extends from the beginning of the Cambrian to the close of the Permian and is marked by the culmination of nearly all classes of invertebrates except the insects and in the later epochs by the appearance of terrestrial plants, amphibians, and reptiles — relating to the corresponding system of rocks—see Chart of Major Events, Section I-2
Pangea	hypothetical land area believed to have once included all present continents
Permian	— the last period of the Paleozoic era or the corresponding system of rocks—see Chart of Major Events, Section I-2
pH	a measure of acidity (lemon juice) and alkalinity (ammonia) of a solution that is a number on a scale on which a value of 7 represents neutrality and lower numbers indicate increasing acidity and higher numbers increasing alkalinity, and on which each unit of change represents a tenfold change in acidity or alkalinity, and that is the negative logarithm of the effective hydrogen-ion concentration or hydrogen-ion activity in gram equivalents per liter of the solution
Photon	a quantum of electromagnetic radiation; they make visible light
Planetesimals	any of numerous small solid celestial bodies that may have existed at an early stage of the development of the solar system
Precambrian	— the earliest era of geological history or the corresponding system of rocks characterized especially by the appearance of single-celled organisms and is equivalent to the Archean and Proterozoic eons—see Chart of Major Events, Section I-2
Prehistoric	the time before written history, about 5,000 years before present
Proton	— an elementary particle that is identical with the nucleus of the hydrogen atom, that along with neutrons is a constituent of all other atomic nuclei, that carries a positive charge numerically equal to the charge of an electron — a particle made of quarks; it forms the nucleus of hydrogen atoms and part of the nuclei of other atoms. (Couper and Henbest, 1997)
Pseudoscience	a system of theories, assumptions, and methods erroneously regarded as scientific
Pumice	— an igneous volcanic rock with a composition like granite and rhyolite and with gas pockets so that it is often sufficiently buoyant to float on water — pumice forms from volcanic froth, or suds
Quarks	The quark family got its name from a quote in James Joyce's book, Finnegans Wake: "Three quarks for Mr. Mark." There are six quarks (three pairs) from heaviest to lightest they are: top and bottom, charm and strange, and up and down. All quarks carry an electric charge: some are positive (such as the up quark) and some negative (such as the down quark). Only the two lightest quarks are stable—the others decay into down and up quarks. Quarks make up protons and neutrons. They are held together by gluons. (Couper and Henbest, 1997)
Quartz	a mineral consisting of silicon dioxide (SiO_2) occurring in various colors (colorless and transparent, yellow, lilac, black, etc.) as hexagonal crystals or in crystalline masses
Quartzite	a compact granular rock composed of quartz and derived from sandstone by metamorphism
Quasar	any of a class of celestial objects that resemble stars but whose large red shift and apparent brightness imply extreme distance and huge energy output
Quaternary	— the geological period from the end of the Tertiary to the present time or the corresponding system of rocks—see Chart of Major Events, Section I-2
Radiation	— the process of emitting radiant energy (for example, heat) in the form of waves or particles — the way in which energy is propagated through the universe; the most familiar form is electromagnetic radiation
Radioactivity	— the emission of energetic particles and/or radiation during radioactive decay — a particular radiation component from a radioactive source, such as alpha, gamma, or beta radioactivity
Radiocarbon Dating	the determination of the age of old material (as an archaeological or paleontological specimen) by means of the content of carbon-14
Rhyolite	an igneous volcanic rock with a composition like granite, and having tiny crystals of quartz and feldspar in a glassy to cryptocrystalline groundmass
Rock	— an aggregate of one or more minerals, e.g. granite, shale, marble; or a body of undifferentiated mineral matter, e.g., obsidian, or of solid organic material, e.g. coal

	— any prominent peak, cliff, or promontory, usually bare when considered as a mass
Rock Cycle	a sequence of events involving the formation, alteration, destruction, and reformation of rocks as a result of such processes as magmatism, erosion, transportation, deposition, lithification, and metamorphism; a possible sequence involves the crystallization of magma to form igneous rocks that are then broken down to sediment as a result of weathering, the sediemnts later being lithified to form sedimentary rocks, which in turn are altered to metamorphic rocks
Sandstone	a sedimentary rock composed of abundant rounded or angular fragments of sand size set in a fine-grained matrix (silt or clay) and cemented with material such as silica, iron oxide, or calcium carbonate); the sand particles usually consist of quartz and the term "sandstone" indicates a rock containing about 85-90% quartz.; may be recognized in the field as a rock containing individual particles that are visible to the unaided eye or slightly larger
Satellite	a manufactured object or vehicle intended to orbit the earth, the moon, or another celestial body
Sediment	— the matter that settles to the bottom of a liquid — material deposited by water, wind, or glaciers
Shale	a fissile (breaks into thin layers) rock that is formed by the consolidation of clay, mud, or silt, has a finely layered structure, and is composed of minerals essentially unaltered since deposition
Siliceous	of, relating to, or containing the element, silicon, for example, window glass is siliceous
Silurian	— the geologic period of the Paleozoic era between the Ordovician and Devonian or the corresponding system of rocks marked by numerous eurypterid crustaceans and the appearance of the first land plants —see Chart of Major Events, Section I-2
Slate	a dense, fine-grained metamorphic rock produced by the compression of various sediments (as clay or shale) so as to develop a characteristic cleavage
Solar Eclipse	the moon passes between the sun and the earth and blocks out the sun's light
Solar System	the sun together with the group of planets (including earth) that are held by its attraction and revolve around it
Striations	several parallel grooves, scratches or channels especially those ground into rocks beneath glaciers
Subduction Zone	a long narrow belt in which one tectonic plate is sliding under an adjacent one; usually also contains a deep trench
Topsoil	surface soil usually including the organic layer in which plants have most of their roots and which the farmer turns over in plowing
Transform Plate Boundary	the point of contact between two tectonic plates that are both squeezing and sliding past one another so that the rocks break with a characteristic pattern of cracks, called transform faults
Tundra	— a level or rolling treeless plain that is characteristic of Arctic and subarctic regions, consists of black mucky soil with a permanently frozen subsoil, and has a dominant vegetation of mosses, lichens, herbs, and dwarf shrubs — a similar region confined to mountainous areas above timberline

How to Convert Between
English and Metric Units of Measurement

	When You Know	*You Can Find*	*If You Multiply By*
Length	inches	millimeters	25
	feet	centimeters	30
	yards	meters	0.9
	miles	kilometers	1.6
	millimeters	inches	0.04
	centimeters	inches	0.4
	meters	yards	1.1
	kilometers	miles	0.6
Area	square inches	square centimeters	6.5
	square feet	square meters	0.09
	square yards	square meters	0.8
	square miles	square kilometers	2.6
	square centimeters	square inches	0.16
	square meters	square yards	1.2
	square kilometers	square miles	0.4
Mass	ounces	grams	28
	pounds	kilograms	0.45
	grams	ounces	0.035
	kilograms	pounds	2.2
Liquid Volume	pints	liters	0.47
	quarts	liters	.95
	gallons	liters	3.8
	liters	pints	2.1
	liters	quarts	1.06
	liters	gallons	0.26
Temperature	degrees Fahrenheit	degrees Celsius	0.556 (after subtracting 32)
	degrees Celsius	degrees Fahrenheit	1.8 (then add 32)

APPENDIX 3

References

Listed Alphabetically By Author

Note: These are mainly the references I used in preparing this guide. For additional information at a K-8 level, search the website: http://www.amazon.com

Bates, R.L. and J.A. Jackson, Eds., 1987. GLOSSARY OF GEOLOGY: American Geological Institute, Alexandria, VA, 788 pp.

Bramwell, Martyn, 1994. ROCKS AND FOSSILS: EDC Publishing Ltd., London, UK, 31 pp.

Brantley, Steven, 1994. VOLCANOES OF THE UNITED STATES: U.S. Dept. of the Interior/U.S. Geological Survey, U.S. Government Printing Office, Washington, DC, 44 pp.

Braun, Esther, and David Braun. THE FIRST PEOPLES OF THE NORTHEAST: Lincoln Historical Society, Lincoln, MA.

Carson, Rachel, 1992. UNDER THE SEA WIND: Truman Talley Books/Plume Printing, New York, NY, 304 pp.

Cole, Joanna, 1987. THE MAGIC SCHOOL BUS: INSIDE THE EARTH: Scholastic, Inc., New York, NY, 40 pp.

Couper, Heather, and Nigel Henbest, 1997. BIG BANG, THE STORY OF THE UNIVERSE, D K Publishing, Inc., New York, NY, 45 pp.

Edwards, Lucy and John Pojeta, 1993. FOSSILS, ROCKS, AND TIME: U.S. Dept. of the Interior/U.S. Geological Survey, U.S. Government Printing Office, Washington, DC, 24 pp.

Fahs, Sophia Lyon, and Dorothy Spoerl, 1960. BEGINNINGS: EARTH, SKY, LIFE, DEATH: Beacon Press, Boston, MA, 217 pp.

Gilman, Richard, C.A. Chapman, T.V. Lowell, and H. W. Borns, Jr.,1988. THE GEOLOGY OF MOUNT DESERT ISLAND, A VISITOR'S GUIDE TO THE GEOLOGY OF ACADIA NATIONAL PARK: U.S. Dept. of the Interior/Maine Geological Survey, Dept. of Conservation, U.S., 50 pp.

Glenn, William, 1975. CONTINENTAL DRIFT AND PLATE TECTONICS: Charles E. Merrill Publishing Company, Columbus, OH, 188 pp.

Hansen, Michael, and Stig Bergstrom, 1997. ANCIENT METEORITES: Ohio Geology Quarterly Publication, Ohio Department of Natural Resources, Division of Geological Survey, Columbus, OH, 32 pp.

Harbaugh, John, 1970. STRATIGRAPHY AND GEOLOGIC TIME: WM. C. Brown Company Publishers, Dubuque, IA, 113 pp.

Kendall, David, 1993. GLACIERS & GRANITE, A GUIDE TO MAINE'S LANDSCAPE & GEOLOGY: North Country Press, Unity, ME, 240 pp.

Matthews, William III, 1967. GEOLOGY MADE SIMPLE: Doubleday & Company, Inc. New York, NY, 192 pp.

McNulty, Faith, 1990. HOW TO DIG A HOLE TO THE OTHER SIDE OF THE WORLD: HarperCollins Publishers, New York, NY, 32 pp.

Mead, William (Foreward), 1995. GEOGRAPHICAL ATLAS OF THE WORLD: Tiger Books International PLC, London, UK, 223 pp.

Newman, William, 1988. GEOLOGICAL TIME: U.S. Dept. of the Interior/U.S. Geological Survey, U.S. Government Printing Office, Washington, DC, 20 pp.

Norton, O. Richard, 1998. ROCKS FROM SPACE, 2nd Ed.: Mountain Press, Missoula, MT, 447pp.

Ohio Department of Natural Resources, Division of Geological Survey. GEOFACTS: A SERIES OF FACT SHEETS ON ASPECTS OF OHIO GEOLOGY, 4383 Fountain Square Drive Columbus, OH 43224-1362. The Ohio Geological Survey offers a rich collection of educational resources—teachers may request a package.

Pinet, Paul, 1992. OCEANOGRAPHY: AN INTRODUCTION TO THE PLANET OCEANUS: West Publishing Company, New York, NY, 572 pp.

Powell, John Wesley, 1978. EXPLORATION OF THE COLORADO RIVER: U.S. Dept. of the Interior/U.S. Geological Survey, U.S. Government Printing Office, Washington, DC, 28 pp.

Press, Frank and Raymond Siever, 1978. EARTH: W.H. Freeman and Company, New York, NY, 649 pp.

Press, Frank and Siever, Raymond, 1998. UNDERSTANDING EARTH: W.H. Freeman and Company, New York, NY, 682 pp.

Renehan, Edward, Jr. 1996. SCIENCE ON THE WEB: A CONNOISSEUR'S GUIDE TO OVER 500 OF THE BEST, MOST USEFUL, AND MOST FUN SCIENCE WEBSITES: Springer-Verlag. New York, Inc., New York, NY, 382 pp.

Rubin, Penni, 1991. MATH IN MOTION: Penni Rubin Publishing, Shaker Heights, OH, 65 pp.

Rubin, Penni, and Eleanora Robbins, 1992. WHAT'S UNDER OUR FEET? U.S. Dept. of the Interior, U.S. Geological Survey: U.S. Government Printing Office, Washington, DC, 42 pp.

Schlee, John, 1991. OUR CHANGING CONTINENT: U.S. Department of the Interior/U.S. Geological Survey, U.S. Government Printing Office, Washington, DC, 21 pp.

Stanley, Steven M., 1998. CHILDREN OF THE ICE AGE: W. H. Freeman and Company, New York, 278 pp.

Stein, Sara, 1986. THE EVOLUTION BOOK: Workman Publishing, New York, 389 pp.

Thompson, Ida, 1996. NATIONAL AUDUBON SOCIETY FIELD GUIDE TO NORTH AMERICAN FOSSILS: Alfred A. Knopf, New York, 846 pp.

Tilling, Robert, C. Heliker, and T. Wright, 1993. ERUPTIONS OF HAWAIIAN VOLCANOES: PAST, PRESENT, AND FUTURE: U.S. Dept. of the Interior/U.S. Geological Survey, U.S. Government Printing Office, Washington, DC, 54 pp.

Tilling, Robert, L. Topinka, and D. Swanson, 1990. ERUPTIONS OF MOUNT ST. HELENS: PAST, PRESENT, AND FUTURE: U.S. Dept. of the Interior/U.S. Geological Survey, U.S. Government Printing Office, Washington, DC, 56 pp.

Tilling, Robert, 1992. VOLCANOES: U.S. Dept. of the Interior/U.S. Geological Survey, U.S. Government Printing Office, Washington, DC, 45 pp.

Van Diver, Bradford, 1997. ROADSIDE GEOLOGY OF NEW YORK: Mountain Press Publishing Company, Missoula, MT, 411 pp.

Van Diver, Bradford, 1995. ROADSIDE GEOLOGY OF VERMONT AND NEW HAMPSHIRE: Mountain Press Publishing Company, Missoula, MT, 230 pp.

Ward, Peter, 1997. THE CALL OF THE DISTANT MAMMOTHS, WHY THE ICE AGE MAMMALS DISAPPEARED: Copernicus, Springer-Verlag New York, Inc., New York, NY, 241 pp.

Watt, Fiona, 1991. PLANET EARTH, A PRACTICAL INTRODUCTION WITH PROJECTS & ACTIVITIES: Usborne Publishing Ltd., London, UK, 48 pp.

Wilber, C. Keith, 1996. THE NEW ENGLAND INDIANS: Globe Pequot Press, Old Saybrook, CT, 88 pp.

Wilson, J. Tuzo (Introductions), 1970. CONTINENTS ADRIFT: W.H. Freeman and Company, New York, NY, 172 pp.

Wright, Thomas, and Thomas Pierson, 1992. LIVING WITH VOLCANOES, THE U.S. GEOLOGICAL SURVEY'S VOLCANO HAZARDS PROGRAM: U.S. Dept. of the Interior/U.S. Geological Survey, U.S. Government Printing Office, Washington, DC, 57 pp.

APPENDIX 4

Websites for General Earth Science Education

Geology Link
•http://www.geologylink.com/
For anyone who has ever been interested in "the world's daily geological rumblings," Geology Link is a "must see" site. You'll find breaking news on geologic events all over the world, the latest news and discoveries, hot topics, virtual field trips, interactive forums, an image gallery and more. From Worth Publishers, this site has something for everyone, from pre-schoolers to professional geologists.

Student Research
•http://www.studyweb.com/science/
General information

Teaching Resources
•http://www.scicentral.com/
Links to many sites where teachers can get information

Geography/Earth Science Education
•http://members.aol.com/bowermanb/101.html

Hands-On Activities for Science Teaching
•http://www.owu.edu/~mggrote/pp/
Provided by Ohio Wesleyan University, Project Primary is a collaborative effort between professors in six departments at the university and K-3 teachers in three Ohio counties to produce hands-on activities for the teaching of science. Activities in Botany, Chemistry, Children's Literature, Geology, Physics, and Zoology, are simply explained and designed to engage children's curiosity and to help them learn. The philosophy of the site is explained in the Constructivism section. Grade Level: Elementary. Content Area(s): Science.

Best Education Web Sites in 1997
•http://www.education-world.com/best_of/1997/reviews.shtml

Teaching about Evolution
•http://www.nap.edu/readingroom/books/evolution98/
The National Academy of Science recently released its new book, *Teaching About Evolution and the Nature of Science*. The teaching of evolution is a controversial issue in many schools.
The publication provides teachers, school administrators, and parents with a framework for helping students

understand this scientific concept. The NSTA's (National Science Teacher's Association) position paper on evolution is included as an appendix in the publication.

Webquest on Earthquakes vs. Volcanoes
•http://www.kn.pacbell.com/wired/fil/pages/webearthscie.html
Here is a new WebQuest from San Diego State's PacBell Fellows. WebQuests are wonderful ways to use the WWW to support an engaging, collaborative, performance task. In this WebQuest, student teams research and debate whether they would rather live near an earthquake fault or a volcano. Students assume the roles of volcanologists and seismologists to gather data. Web links help students learn earth science principles as they consider their task. Students finish the quest by writing e-mail to a geologist. This site aligns nicely with the Earth, Inquiry and Problem Solving, Reasoning, and Communication Standards.

WebQuest on Cloning
•http://204.102.137.135/PUSDRBHS/science/clone/hello.htm
Check out the matrix of examples. I think "Hello Dolly" (social and ethical implications of cloning) is one of the best examples of a WebQuest I have seen. It begins with an actual performance task and involves students in working together as a team with assigned roles and ends with a shared presentation.
For those of you interested in learning more about WebQuests, visit San Diego State University's WebQuest site:
•http://edweb.sdsu.edu/webquest/webquest.html

Science at the Movies
•http://www.scienceweb.org/movies/nowshowing.html
Interest your students in the scientific concepts behind many of the special effects found in a number of popular movies at the web site Science Web Goes to the Movies. Movies include *The Abyss*, *Stargate*, and *Speed*. Students are also invited to submit science-related questions about any movie.

Transition to Standards-Based Learning
•http://www.nap.edu/bookstore
Every Child a Scientist: Achieving Scientific Literacy for All is the newest publication from the National Research Council. The 26-page booklet offers guidance to parents and others on how to help their local schools make the transition to standards-based teaching and learning. For more information call 1-800-624-6242.

Understanding Our Planet Through Chemistry
•http://helios.cr.usgs.gov/gips/aii-home.htm
This U.S.Geological Survey site shows how chemists and geologists use analytical chemistry to: determine the age of the earth; show that an extraterrestrial body collided with the earth; predict volcanic eruptions; observe atmospheric change over millions of years; and document damage by acid rain and pollution of the earth's surface.

More Earth Science Links
•http://meguma.earthsciences.dal.ca/~dawson/link.html

Evolution and the Teaching of Science.
•http://www.nap.edu/readingroom/books/evolution98/
The National Academy of Sciences has just released a new publication, "Teaching About Evolution and the Nature of Science." The publication which can be perused at this site provides teachers, school administrators, and parents with a framework for helping students understand this scientific concept.

About Hawaiian Volcanoes
•http://hvo.wr.usgs.gov/
This USGS site from the Hawaiian Volcano Observatory offers something for everyone. From fun activities for kids to learn about volcanoes to a volcano watch for the latest Hawaiian volcanic eruptions, to how volcanoes work.